MATHEMATICAL
CONVERSATIONS

MATHEMATICAL CONVERSATIONS

Multicolor Problems,
Problems in the Theory of Numbers,
and *Random Walks*

E. B. Dynkin
and
V. A. Uspenskii

DOVER PUBLICATIONS, INC.
Mineola, New York

Bibliographical Note

This Dover edition, first published in 2006, is an unabridged republication and consolidation of the three booklets originally published by D. C. Heath, Boston, in 1963: *Multicolor Problems* (translated and adapted, from the Russian, by Norman D. Whaland, Jr., and Robert D. Brown), *Problems in the Theory of Numbers* (translated and adapted by Norman D. Whaland, Jr., and Michael B. P. Slater), and *Random Walks* (translated and adapted by Norman D. Whaland, Jr., and Olga A. Titelbaum). The original three-part Russian work (*Matematicheskie besedy: zadachi o mnogotsvetnoi raskraske, zadachi iz teorii chisel, sluchainye bluzhdaniya*) was published by Gosudarstvennoe izdatel'stvo tekhniko-teoreticheskoi literatury, Moscow, in 1952.

Library of Congress Cataloging-in-Publication Data

Dynkin, E. B. (Evgenii Borisovich), 1924-
  [Matematicheskie besedy. English]
  Mathematical conversations : Multicolor problems, Problems in the theory of numbers, and Random walks / E.B. Dynkin and V.A. Uspenskii.
      p. cm.
  "This Dover edition . . . is an unabridged republication and consolidation of the three [translated and adapted] booklets originally published by D.C. Heath, Boston, in 1963"—T.p. verso.
  The original three-part Russian work, Matematicheskie besedy, was published in Moscow in 1952.
  Includes bibliographical references.
  ISBN 0-486-45351-0 (pbk.)
  1. Mathematics—Problems, exercises, etc. I. Uspenskii, V. A. (Vladimir Andreevich) II. Title.

QA43.D913 2006
510—dc22

                                                                    2006050215

Manufactured in the United States of America
Dover Publications, Inc., 31 East 2nd Street, Mineola, N.Y. 11501

PREFACE TO THE AMERICAN EDITION

Mathematical Conversations by E. B. Dynkin and V. A. Uspenskii (published as Number 6 in the Russian series, *Library of the Mathematics Circle*) was based on the material covered during the academic years 1945–1946 and 1946–1947 in one of the sections of the School Mathematics Circle at Moscow State University. One of the authors was the instructor of the section, and the other was a participant.

The primary aim was not so much to impart new information as to teach an active, creative attitude toward mathematics. The most successful topics took shape only as the work progressed. A series of consecutive meetings was devoted to each topic. A meeting would usually begin with problems whose formulation required no new concepts, but whose solution led the students directly into a new area of inquiry. These problems would sometimes be solved during the meeting, but more often they were left as homework. At the next meeting, the instructor would discuss the solutions of the problems and then use them as a basis for generalizations. Whenever possible, the material was presented in sequences of related problems.

This book presents three of the topics in a considerably revised and expanded form. Like the discussions on which it is based, it retains the practice of interrupting the presentation with problems whose solutions are essential to what follows. To understand this material, the reader should be familiar with high-school algebra.

INSTRUCTIONS FOR THE USE OF THIS BOOK

Each section of this book is devoted to a single topic, and therefore can be read separately. Each section is designed for the reader's active participation, and the associated problems form an organic part of the text. Most of the problems are grouped in sequences, each sequence forming a unit and building up to a final result contained in the last problem of the sequence (in Section I, for example, Problems 21–27 and 38–41; in Section II, Problems 28–32 and 71–75). Sometimes the aim of a sequence of problems is not some definite result, but rather mastery of a new method (in Section I, Problems 11–14; in Section II, Problems 56–58). Finally, a few of the problems are practice exercises, designed to help the reader master new concepts (in Section II, for example, Problems 1–8; in Section III, Problems 1–3).

Before attempting to solve a problem, the reader should examine all the problems in the given sequence. Solutions are provided at the end of each section, but it is recommended that the reader look at them *only* after he has tried to solve *all* problems of a sequence. If he looks at the solutions too soon, they may set his mind working in a certain direction, but with independent thought he may arrive at new and original methods. The experience of the School Mathematics Circles has shown that sometimes simpler and more elegant solutions are found than those expected by the authors of the problems.

The reader may not always be able to solve all the problems of a sequence independently. If, after solving the first few problems, he should run into difficulties, he may find it helpful to read the solutions of the problems he has already solved. If these do not suggest an approach to the next problem, he should look at its solution, and then proceed to try the rest by himself. Eventually, he should read all the solutions, whether or not he has succeeded in solving the problems independently, as they have been carefully prepared, and many of them are accompanied by conclusions and remarks of a fundamental nature.

Although the problems here are basic, this is by no means merely an exercise book. The text is also important. The relation between problems and text differs in the various chapters. Sometimes the essential ideas are set forth in the text, but in others they are contained in the problems, and the text merely introduces concepts and states results. The text and problems are always closely related and must be read in the order in which they appear in the book.

In conclusion, we advise the reader not to begrudge the time spent on solving the problems. Each sequence, indeed each problem, solved independently enlarges the arsenal of resources at his disposal. One idea arrived at independently is worth a dozen borrowed ones. Even if persistent attempts to solve a problem do not lead to success, the time is not spent in vain, as he will then see its solution in a new light. He can look for the reason for his failure and can discover the fundamental idea that leads to success.

CONTENTS

SECTION ONE
Multicolor Problems

Note

THIS section is devoted to a single topic, and the main part of it should therefore be read in order. But the Appendix consists of the solution of a related problem and can be read independently of the rest of the section, since only the definition of *proper coloring* is necessary to understand it.

To understand most of the rest of the material in this section, the reader should be familiar with high-school algebra and with the method of mathematical induction. See, for example, *The Method of Mathematical Induction* by I. S. Sominskii (Boston: D. C. Heath and Company, 1963).

Readers will find a great deal of useful and entertaining information that updates this section in *Four Colors Suffice: How the Map Problem Was Solved* by Robin Wilson (Princeton, New Jersey: Princeton University Press, 2004).

Introduction

On a geographical map different regions or countries are, for convenience, colored with different colors. Ordinarily, however, each region need not have its own separate color. It is sufficient to color with different colors only *neighboring* regions, that is, regions having a common boundary, such as regions S_1 and S_2 in Figure 1.[1]

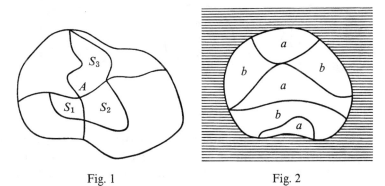

Fig. 1   Fig. 2

We shall use the following definition:

DEFINITION. *A* coloring of a map *is defined as* proper *if neighboring regions are colored with different colors.*

It is natural to ask, how many colors are needed to color a given map *properly?* Obviously, one answer is to use just as many colors as there are regions; in this case, we simply color each region with its own color. But we can hardly be satisfied with the solution. We are interested in the *minimum* number of colors sufficient for a proper coloring of a given map. It is easy to construct a map for which this minimum number of colors is two (Fig. 2).

The map in Figure 2 is the map of an island. The sea (shown by shading) surrounding the island is colored with neither color *a* nor color *b*. Usually, however, the sea is also colored on maps, so that

---

[1] The regions S_1 and S_3 are not considered to be neighboring. Although they come into contact at point A, they have no common *boundary.*

regions with a seacoast, that is, regions that border on the sea, must be colored differently from the sea. Consequently, for our purposes the sea is no different from an ordinary region. It does not matter that the sea is unbounded. Hereafter, therefore, we shall not consider the sea as separate, but shall instead include it among the other regions. Thus, the maps we shall consider hereafter will not be maps of islands, but will be thought of as extending over the entire plane. From this point of view, the map in Figure 2 can no longer be properly colored with two colors.

Let us now return to the problem of the minimum number of colors sufficient to color a map properly. In Figure 3 maps are rep-

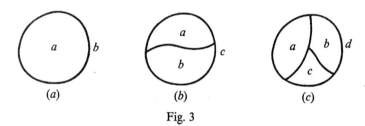

(a) $\qquad$ (b) $\qquad$ (c)

Fig. 3

resented for which these numbers are two, three, and four, respectively. But here our examples come abruptly to an end. Up to now, no one has been able to construct a map for which the minimum number of colors is five or more, in other words, which cannot be properly colored with four colors. It has been assumed that every map can be properly colored with four colors, but no one has yet proved it. (This is the famous *four-color problem*.) It has been proved, on the other hand, that every map can be properly colored with five colors. (This we shall prove in Chapter 4.) We are therefore able to make only the following two assertions, and we cannot fill the disappointing gap between them:

(1) *Not every* map can be properly colored with *three* colors (see Fig. 3c).

(2) *Every* map can be properly colored with *five* colors.

In the following chapters, we shall be concerned with questions about maps for which two colors (Chapter 1) and for which three colors (Chapter 2) are sufficient. In Chapter 3 we shall seek to deduce some criteria for coloring with four colors; in Chapter 4 we shall prove the *five-color theorem*.

1. Coloring with Two Colors

1. SIMPLE TWO-COLOR PROBLEMS

Problem 1.[1] Let n straight lines be drawn in the plane. Prove that the map formed by them can be properly colored with two colors (Fig. 4).[2]

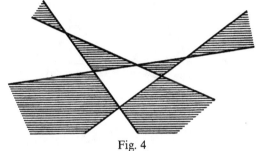

Fig. 4

Problem 2. Let n circles be drawn in the plane. Prove that the map formed by these circles can be properly colored with two colors (Fig. 5).

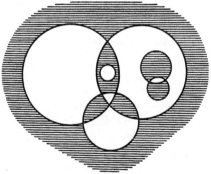

Fig. 5

---

[1] See the Preface for Instructions for the Use of This Book.

[2] *Hint.* A single straight line divides a plane into two parts; two parallel straight lines divide it into three parts; two intersecting straight lines divide it into four parts. All these maps can be properly colored with two colors. A proof can now be completed by mathematical induction.

DEFINITION. *If a plane is partitioned into triangles in such a way that any two triangles either have no common point, or have a common vertex, or have a common side, such a partition is called a* triangulation.[1]

DEFINITION. *If at the vertices of a triangulation the digits 0, 1, and 2 are placed in such a way that the vertices at the two ends of the same side are numbered with different digits (Fig. 6), such a* numbering of vertices *is called* proper.

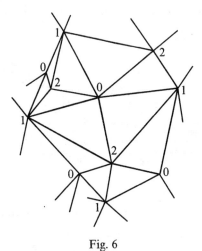

Fig. 6

Problem 3. Prove that a triangulation with proper numbering of vertices can be colored with two colors, assuming that it extends indefinitely over the plane.

Problem 4. If on a map there is a region for which the number of boundaries is not divisible by m, while the number of boundaries of each other region is divisible by m, then the map cannot be properly colored with two colors.

[1] Figure 7 shows examples of partitions into triangles that are not triangulations.

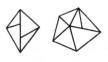

Fig. 7

2. PROBLEMS ON SQUARE BOARDS

The customary coloring of the squares of a chessboard serves as an example of a proper coloring provided we disregard the region outside. The problems on the chessboard that we introduce here will help us later on to solve the general two-color problem.

A knight can, under the rules of chess, go in one move from the square S to any one of the squares S_1–S_8 (Fig. 8). A rook can go in

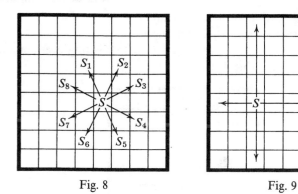

Fig. 8   Fig. 9

one move from square S to any square in the corresponding row or column (Fig. 9). In solving problems, we agree to regard a rook going from a square S to a square S' (Fig. 9) as visiting all the intervening squares as well.

Problem 5. With a knight, make the rounds of all the squares of a "chessboard" consisting of 5 × 5 squares in such a way that the knight lands on no square twice.

Problem 6. Number the squares of a 25-square "chessboard" in the order in which the knight rested on them in the previous problem. Shade all the squares that have an even number. Show what coloring results if the same operations are carried out, not on a 25-square "chessboard," but on one with an arbitrary number of squares, which a knight can visit as indicated in Problem 5.

Problem 7. Is it possible for a knight to touch once each of the squares of a 49-square board and on the last move to come to a square neighboring the square from which he started?

Problem 8. Prove that in one circuit a knight cannot visit all the squares of a 49-square board if he starts from the square *S* (Fig. 10).

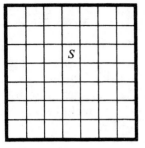

Problem 9. A knight has made *n* moves and has returned to the square from which he started. Prove that *n* is even.

Fig. 10

Problem 10. Prove that a rook cannot move from corner *A* of a 64-square chessboard to the diagonally opposite corner *B* by visiting every square once and only once (Fig. 11).

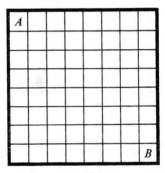

Fig. 11

3. PROBLEMS INVOLVING EVEN AND ODD NUMBERS

Problem 11. Can one arrange all 28 dominoes of a (double-six) set in a single chain so that the number 6 is at one end and the number 5 at the other?

Problem 12. Every human being who ever lived on the earth has, in the course of his life, shaken hands some completely definite number of times. Show that the number of human beings who have shaken hands an *odd* number of times must be *even*.[1]

Problem 13. At a meeting, 225 persons were present. Friends shook hands with each other. Prove that at least one of the participants present at the meeting shook hands with an *even* number of people.

---

[1] Zero is an even number; hence, a human being who has never shaken hands has shaken hands an even number of times.

Problem 14. In Figure 12, six points A, B, C, D, E, F are shown, and each of these points is connected with three of the remaining

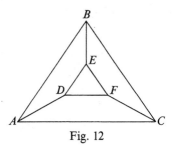

Fig. 12

five. Prove that if instead of six there are given only five points, then it is impossible to draw curves connecting them in such a way that each of the points is connected to exactly three of the remaining points.

4. NETWORKS AND MAPS

We shall now define several terms that we have been using intuitively up to now. Before doing so, let us consider an arbitrary network of curves in the plane and introduce a new term.

DEFINITION. *If from a particular point of this network one can move away along curves of the network in k different directions, then we shall say that the* multiplicity *of this point is k.*

For example, for the network represented in Figure 13, the multiplicity of the point A is 1, the multiplicities of the points B and

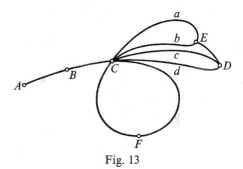

Fig. 13

F are each 2, the multiplicities of the points D and E are each 3, and, finally, the multiplicity of the point C is 7.

DEFINITION. *Any point of the network whose multiplicity is different from 2 is a* vertex *of the network.*

The network of Figure 13 has, by this definition, four *vertices, A, C, D,* and *E.*

DEFINITION. *The arc of any curve of the network between two adjacent vertices is called a* boundary.

In this way, each *boundary* contains two vertices. (In special cases, these two vertices can coincide to become one.) In our network (Fig. 13), there are seven boundaries: *ABC, ED, CaE, CbE, CcD, CdD,* and *CFC.* In the last of these, the two vertices containing between them the boundary *CFC* coincide. We shall denote the number of vertices of a network by v, and the number of boundaries by b.

Problem 15. Construct a network of curves for which

$$(a) \ v = 3, \ b = 5;$$
$$(b) \ v = 7, \ b = 11.$$

Problem 16. Let a network of curves have b boundaries and v vertices with the multiplicities $k_1, k_2, \ldots, k_v$. Prove that

$$k_1 + k_2 + \cdots + k_v = 2b.$$

Problem 17. Prove that in any network of curves the number of vertices having odd multiplicities is even.

Notice that not every network of curves can be called a *map.*

DEFINITION. *A* map *is a network such that every* boundary *must necessarily separate two neighboring* regions.

Hence, for example, there can be no vertices of multiplicity 1 on a *map.* In Figure 13, the boundary *ABC*, which goes out from the vertex *A* of multiplicity 1, does not separate any two *regions.* We shall denote the number of regions of a map by r and count the outside region along with the others.

Problem 18. Construct a map for which

$$(a) \ v = 5, \quad b = 8, \quad r = 5;$$
$$(b) \ v = 11, \quad b = 19, \quad r = 10;$$
$$(c) \ v = 6, \quad b = 12, \quad r = 9.$$

Problem 19. A map has b boundaries and r regions, which have $n_1, n_2, \ldots, n_r$ boundaries, respectively. Prove that

$$n_1 + n_2 + \cdots + n_r = 2b.$$

Problem 20. Prove that, in an arbitrary map, the number of regions having an odd number of boundaries is even.

Note. We have agreed to call *vertices* those points of a network of curves having a multiplicity different from 2. Sometimes, however, it is convenient to consider as vertices also certain points of multiplicity 2. As before, a boundary is a section of any curve lying between two successive vertices. For example, the map in Figure 14 has 9 vertices, $A, B, C, D, E, F, G, H, I$, and 11 boundaries, $AB, BC, CD, DE, EG, GF, FA, BI, IH, HD, HG$. It is easy to verify that the statements and solutions of all the problems formulated earlier are still valid under the new meaning of *vertex*.

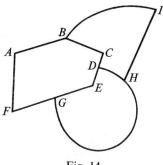

Fig. 14

5. GENERAL TWO-COLOR PROBLEMS

By analogy with the chessboard, we now introduce a rook on an arbitrary map. The rook goes through the various regions, being able to go in one move from any region into any neighboring one (in Fig. 15, from S to any one of S_1, S_2, S_3, S_4, S_5).

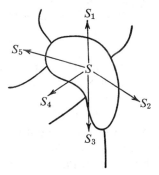

Fig. 15

Problem 21. With a rook, make a tour of all the regions of the map represented in Figure 16 without visiting any region twice. Number the regions in the order in which the rook visited them, and shade those regions that are given an even number.

Fig. 16

Problem 22. Prove that it is impossible for a rook to tour all the regions of the map represented in Figure 17 without visiting at least one region twice.

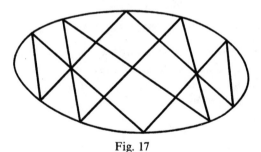

Fig. 17

Problem 23. Let a map be properly colored with two colors. Prove that all of its vertices have an even multiplicity.

Problem 24. Suppose all of the vertices of a map have an even multiplicity. A rook tours a series of regions (not necessarily all) of this map, without visiting any one of them twice, and returns to the region from which it started. Prove that the rook has made an even number of moves.

Problem 25. As in Problem 24, suppose all of the vertices of a map have an even multiplicity. A rook tours a series of regions of this map and returns to the region from which it started. (This time, in the process, the rook could have visited some regions more than once.) Prove that the rook has made an even number of moves.

Problem 26. Let all vertices of a map have an even multiplicity. Let a rook go along a certain path from the region S_0 to the region S_1 in p moves, and along another path in q moves. Prove that the numbers p and q are either both even or both odd.

Problem 27. Let all vertices of a map have an even multiplicity, including 2. Prove that the map can be properly colored with two colors (compare with Problem 23).

Problems 23 and 27 yield the following theorem, which completely solves the problem of the proper coloring of a map with two colors:

THEOREM. *A map can be properly colored with two colors if and only if all of its vertices have an even multiplicity, including 2.*

2. Coloring with Three Colors

6. A SIMPLE THREE-COLOR PROBLEM

Problem 28. Suppose n circles are drawn in a plane. In each circle let a chord be drawn so that chords of two different circles have at most one point in common. Prove that a map obtained in this way can always be properly colored with three colors. (Figure 18 is an example of such a map.)

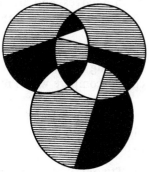

Fig. 18

7. PROBLEMS ON HEXAGONAL BOARDS

The hexagonal board pictured in Figure 19 has the same significance for the three-color problem that the chessboard has for the two-color problem. Unlike the ordinary chessboard, it is not composed of squares, but rather of regular hexagons, and can be properly colored with three colors, say white, black, and red[1] (Fig. 20), provided we disregard the region outside. For such a board, we could devise rules of play analogous to those for ordinary chess. We limit ourselves to introducing a playing piece that we shall call a *camel*.[2] In one move, the camel can go from a region in any of the three directions shown by arrows in Figure 19: up, down to the left, or down to the right; that is, it can go from region S to any one of the regions S_1, S_2, or S_3. In Figure 20, the path of a camel from the bottom corner of the board to the top, and from the top to the bottom is shown.

Our hexagonal board itself has the form of a hexagon. An "edge" of this large hexagon, or a "side" of the board, may consist of any number of hexagonal regions. (In Figure 19 a side of the board consists of five hexagonal regions.)

---

[1] In Figure 20, red is indicated by horizontal shading.
[2] The piece received this name in the school mathematics circles in Russia.

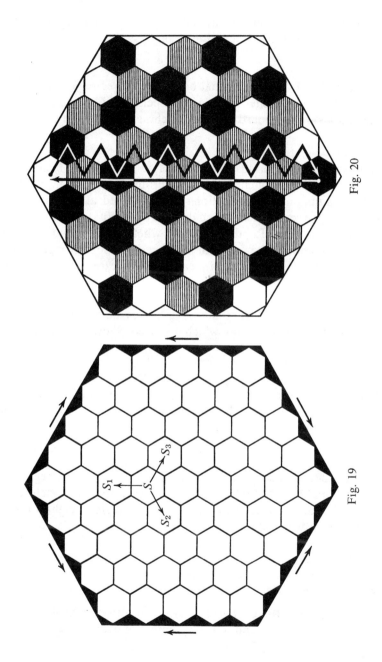

Fig. 20

Fig. 19

Problem 29. Determine the number of regions in a hexagonal board whose sides have five, six, and m hexagons.

Problem 30. Draw a hexagonal board with three regions on a side. Find a path for a camel starting in the center region and visiting all the regions on the board without touching any region twice.

Problem 31. Number all of the regions on the board in the order in which the camel touched them in Problem 30. Color black all regions numbered with multiples of 3, and red all regions whose numbers give remainder 1 when divided by 3. What kind of coloring results? What coloring do we obtain when we carry out the same process on another board, with a different number of regions on a side, that can be toured with a camel as indicated in Problem 30?

Problem 32. Suppose that a camel has made n moves and has returned to the region from which it started. Show that n is divisible by 3.

Problem 33. Prove that if a camel starts out from a corner region, it cannot tour a hexagonal board with three regions on a side, visiting every region only once.

Problem 34. Is it possible for a camel to tour all the regions of a hexagonal board with m regions on a side, touching each region only once, and arrive, on the last move, in a region neighboring the region from which it started?

8. DUAL DIAGRAMS

Let us mark the center of each region of the hexagonal board shown in Figure 19 and join the centers of each pair of adjoining regions by a line segment. If we erase the edges of the hexagonal regions, leaving only the centers and the line segments that connect them, we obtain the diagram in Figure 21. Points in Figure 21 correspond to regions of the board in Figure 19. The diagram we have drawn is very convenient for solving problems involving the path of a camel. (The directions in which a camel can move are shown by arrows.)

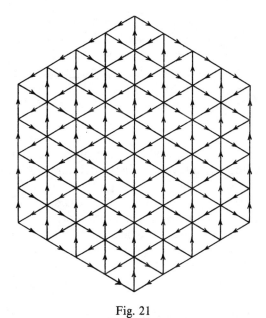

Fig. 21

Problem 35. Prove that it is impossible for a camel to tour all the regions of the hexagonal board in Figure 19 (or, what amounts to the same thing, all the points of the diagram in Figure 21) without visiting some region twice.

DEFINITION. *Diagrams related as are Figures 19 and 21 are* dual *to each other.*

A system of 25 points, connected by straight lines with arrows, is shown in Figure 22. A board *dual* to this diagram is shown

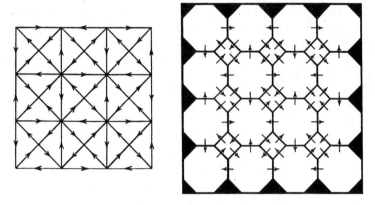

Fig. 22   Fig. 23

in Figure 23. This board is composed of regions of two kinds, octagons and squares. If we mark each octagon and each square at the center, join the centers of neighboring regions, and erase the boundaries of the regions, we obtain the diagram in Figure 22 again.

Problem 36. Devise a camel's tour of the diagram in Figure 22 so that no point is visited twice. Number the points of the diagram in the order in which they were touched and replace each number by its remainder when divided by 3. Color black all regions of the board in Figure 23 that correspond to points numbered 0 in Figure 22 and color red all regions corresponding to points numbered 1.

Problem 37. Prove that it is impossible for a camel to tour all of the regions of the board in Figure 23, touching each region only once, if it starts its tour in an octagonal region.

9. TRIANGULATION

We can consider the paths of camels on diagrams considerably more general than the diagrams shown in Figures 21 and 22. We recall that by a *triangulation* of a polygon we mean a decomposition into triangles, any two of which have either a vertex, a side, or no point at all in common (see first definition in section 1). Let us assume that some polygon (or even the whole plane) is triangulated, and that the triangles into which it (or the plane) is partitioned are properly colored with two colors, white and black (see, for example, Figure 24).[1] A playing piece that moves along the sides and vertices of the triangles, going in one move from any vertex to one of its *neighboring* vertices (that is, end points of the same side), will again be called a *camel*. Furthermore, the direction of motion will be such that, as the camel moves along a side, a black triangle always lies to the right and a white one to the left. In Figure 24 the possible directions of a camel's motion are indicated by arrows.

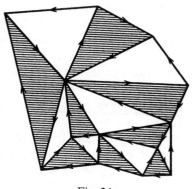

Fig. 24

Instead of referring to a special diagram, the following four problems, unlike the previous groups of problems, relate to arbitrary diagrams of the sort described.

Problem 38. Prove that a camel can go from any vertex of a diagram to any other.

Problem 39. Suppose a camel takes n steps and returns to its starting place. Prove that n is divisible by 3 (see Problem 4).

Problem 40. Let A and B be two arbitrary vertices of a diagram. A camel can go from point A to point B in different ways. Let p be the number of moves in one of these ways and q the number of moves in another (Fig. 25). Prove that $p - q$ is divisible by 3.

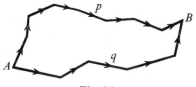

Fig. 25

---

[1] Notice that here the outside (sea) is not one of the regions by the definition of triangulation. It cannot be either white or black. However, if we annex a black triangle to each white one bordering on the "sea," then the "sea" can be white even though it is not a "triangle."

Problem 41. Let a polygon or the plane be triangulated, and let the triangles of the division be properly colored with two colors. Prove that the vertices of all of the triangles can be numbered with the digits 0, 1, and 2 in such a way that any two neighboring vertices have different digits (compare with Problem 3).[1]

The number of colors necessary for a proper coloring of a map is obviously not in the least dependent on the size of the regions or on the shape and length of the boundaries. It is determined only by the relative positions of the regions, boundaries, and vertices. If we draw a map on a sheet of rubber and stretch this sheet unequally (without ripping it), all maps thereby obtained from the original map are completely equivalent to one another from our point of

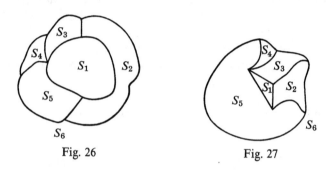

Fig. 26   Fig. 27

view. For example, the two maps in Figures 26 and 27 are equivalent.

In the formulation of Problems 3 and 41, the maps were said to be made from triangles with straight sides. The solution of these problems is unchanged, however, if we consider maps with curved sides instead of straight ones. Hence, the result of Problem 41 can be formulated in the following way:

THEOREM. *Let a map consist of regions which have three boundaries each.[2] If this map can be properly colored with two colors, then its vertices can be properly numbered with three digits.*

The result of Problem 3 can be generalized in the same way. The solution remains exactly the same, since two neighboring triangles

---

[1] Recall our definition of *proper* numbering of vertices in section 1.

[2] We shall call such regions *triangles;* in general, we shall call a region with *n* boundaries an *n-gon*.

have, as before, different orientations, although they can now border one another in two different ways (Fig. 28*a* and *b*).

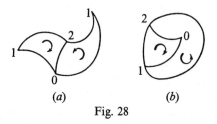

(*a*)   (*b*)

Fig. 28

10. DUAL MAPS

Consider the map in Figure 29. Inside each of its regions mark a point, the capital of the region. Let us connect the capitals of each pair of neighboring regions by a railroad (dotted lines in Figure 29),

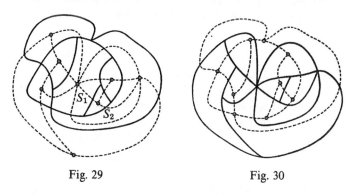

Fig. 29   Fig. 30

which does not go outside these two regions nor through any vertex. If any two regions have several boundaries in common, as, for example, the regions S_1 and S_2 in Figure 29, we join the two capitals by several railroads, one through each common boundary. In doing this, we take care to see that two different railroads do not cross one another.

If we replace the dotted lines on our map by solid lines, and the solid lines by dotted lines, we obtain the map shown in Figure 30. The original map and the map of the railroad lines exchange roles; the original map is the railroad map for its own railroad map. Consequently, the two maps in Figure 29 (or in Fig. 30), one drawn with dotted lines, the other with solid lines, play a completely symmetric role—each of them is the railroad map for the other.

We shall call two such maps *dual maps,*[1] defined as follows:

DEFINITION. *A map is said to be the* dual *of a second map if the two maps play completely symmetric roles, one to the other, under the following conditions:*

(1) *Each boundary of one map intersects exactly one boundary of its dual map.*

(2) *In the interior of every region of one map, there is exactly one vertex of its dual map. (In this way, a one-to-one correspondence is set up between the elements (regions, boundaries, vertices) of two dual maps,[2] such that regions of one map correspond to vertices of the dual, vertices correspond to regions, and boundaries to boundaries.)*

(3) *Neighboring regions of one map correspond to neighboring vertices of the dual map, and vice versa.*

(4) *If one of the vertices of one map has multiplicity k, then the region in the dual map corresponding to this vertex has k boundaries; that is, according to our convention, it is a k-gon.*

If we now try to construct railroad lines for arbitrary maps, we encounter difficulties of two kinds.

In the first place, it is possible that the diagrams of railroads do not represent maps at all in our sense. We consider, for example, the diagram of railroads for the map shown in Figure 31 (this diagram is drawn separately in Figure 32). In this diagram, there exist

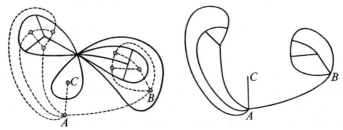

Fig. 31   Fig. 32

[1] The construction by which we obtained the map in Figure 21 from the map in Figure 19 differs from the method we have just used only in that there we did not take the exterior region into account. Hence, to obtain the map dual to the map in Figure 19, it is sufficient to add a single vertex to the map in Figure 21, joining it to all of the outside vertices of that map. The same can be said of the maps shown in Figures 22 and 23.

[2] That is, each element in one map corresponds to a single well-determined element of the dual map.

"boundaries" (*AB* and *AC*) which do not separate regions; one and the same region lies on both sides of each of these boundaries.

Also, the diagram of railroads for the map shown in Figure 33 represents a true map, but the exterior region of the railroad map

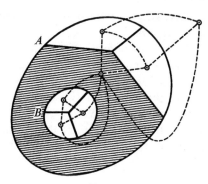

Fig. 33

contains two vertices, *A* and *B*, of the original map. In this case, there is no duality between the original map and its railroad map.

The second difficulty can be avoided if we consider only *connected* maps.

DEFINITION. *A map is* connected *if no one region separates the remaining regions into two or more groups that are nowhere contiguous.*

In other words, if we draw a connected map on a sheet of paper, then, no matter what region we cut out with a pair of scissors, the remaining part does not fall into separate pieces. The map shown in Figure 33 is not a connected map. (The "separating" region is shaded.) A connected map can also be defined as a map in which any given vertex can be reached from any other vertex by moving along the boundaries of the map.

We could overcome the first difficulty in the same way as the second by simply excluding from consideration maps for which this difficulty arises. But instead we shall introduce maps similar to the "maps" in Figure 32, that is, maps with boundaries that do not separate regions. In what follows, we shall almost never encounter such maps; they are introduced here only so that our *duality principle* is not violated.

We now formulate the *duality principle* for connected maps.

(*a*) The railroad map of a connected map is again a map; moreover, it is a connected map.

(*b*) Every connected map is the railroad map of its own railroad map.

Hence, a connected map and the corresponding railroad map are dual to each other, and the important relationships between the duals we named earlier hold here also.

However, for condition 4 on page 20 to remain true, we need the following modification: every boundary not separating regions must be counted twice when computing the number of boundaries of a region. For example, the region in Figure 34 is an octagon (the multiplicity of the corresponding vertex A of the dual map is 8). With this stipulation, the results of all of the problems in which the number of boundaries of regions was to be determined remain true, as, for example, Problems 19 and 20.

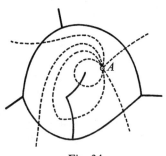

Fig. 34

We can now state the following result, which follows directly from condition 3 of dual maps:

THEOREM. *If the regions of a map can be properly colored with n colors, then the vertices of the dual map can be properly numbered with n numbers. Conversely, if the vertices of a map can be properly numbered with n numbers, the regions of the dual map can be properly colored with n colors.*[1]

---

[1] Notice that it is meaningless to speak of a proper coloring of a "map" containing boundaries which do not separate regions. In fact, in such a map there is a region that lies on both sides of one of its boundaries, that is, a region neighboring itself. See the Definition in section 4.

11. NORMAL MAPS IN THREE COLORS

DEFINITION. *A* normal map *is a map all of whose vertices have multiplicity 3.*

Figure 35 shows an example of a normal map. The significance of normal maps will be made clear in Chapter 3.

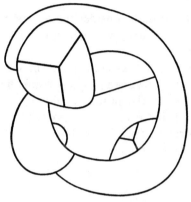

Fig. 35

Problem 42. Prove that a normal map can be properly colored with three colors if and only if the number of boundaries of each of its regions is even.

It should be noted that the theorem formulated in Problem 42 does not completely solve the problem of proper coloring with three colors, since it holds only for normal maps.

3. The Four-Color Problem

12. NORMAL MAPS IN FOUR COLORS

In this chapter, we shall consider the question of coloring maps with four colors. (Of course, we shall not succeed in settling this question completely.)

In solving the general four-color problem, it is sufficient to consider *normal maps;* for, if we can properly color all normal maps with four colors, then we can also properly color all maps in general with four colors. Indeed, if we have a vertex with multiplicity greater than 3 in an arbitrary map, we can draw a small circle around the vertex, remove all of the boundaries in the interior of the circle, and adjoin its interior to one of the adjacent regions (Fig. 36). We then

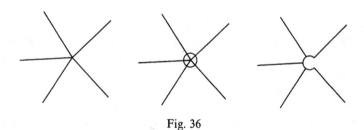

Fig. 36

obtain some normal map. If this map can be properly colored with four colors, this will obviously be true of the original map also. Hence, in what follows we shall consider only normal maps.

Up until now we have considered only proper colorings of the regions and proper numberings of the vertices of maps. We now introduce proper numbering of boundaries. Two boundaries are said to be *neighboring* if they have a vertex in common.

DEFINITION. *A numbering of boundaries is called proper if any two neighboring boundaries have different numbers.*

Figure 37 shows an example of a *proper* numbering of boundaries.

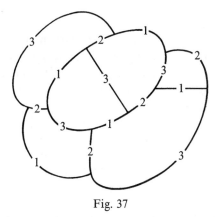

Fig. 37

13. VOLYNSKII'S THEOREM

The four-color problem is equivalent to the problem of properly numbering boundaries with three digits. This equivalence is established by the following theorem.

VOLYNSKII'S THEOREM.[1] *A normal map can be properly colored with four colors if and only if its boundaries can be properly numbered with three digits.*

The proof of this theorem follows from Problems 43 to 45. In the following, it will be convenient to number regions with four numbers rather than to color them with four colors. For this purpose we choose the number-pairs (0,0), (0,1), (1,0), (1,1). These pairs can be added termwise, and, in order to use no other number-pair in the process, we replace sums by their remainders after dividing by 2. For example,

$$(1,0) + (1,0) = (0,0),$$
$$(0,1) + (1,1) = (1,0).$$

[1] V. V. Volynskii (1923–1943) was a talented young mathematician, a student of the department of mechanics and mathematics of Moscow State University. He fell at the front in World War II.

Volynskii's Theorem was actually discovered in the nineteenth century by the Scottish physicist and mathematician P. G. Tait (1831–1901). See P. G. Tait, "Remarks on the Colouring of Graphs," *Proceedings of the Royal Society of Edinburgh,* vol. 10 (1880), p. 729.

Problem 43. Prove that if a normal map can be properly colored with four colors, then its boundaries can be properly numbered with three digits.

Problem 44. Let the boundaries of a normal map be properly numbered with the three number-pairs (0,1), (1,0), (1,1). A rook visits a series of regions of this map and returns to its starting point. Prove that the sum of the numbers of all the boundaries crossed by the rook is equal to (0,0).

Problem 45. Let the boundaries of a normal map be properly numbered with three number-pairs. Prove that this map can be properly colored with four colors. (This completes the proof of Volynskii's theorem.)

Problem 46. Let the number of boundaries of each region of a normal map be divisible by 3. Using Volynskii's theorem, prove that this map can be properly colored with four colors.

At the end of the following section, after the reader will have learned about Euler's theorem, the four-color problem will be solved for maps with fewer than twelve regions.

4. The Five-Color Theorem

14. EULER'S THEOREM

The boundaries of some region can fall into separate, disconnected contours (Figs. 38a and b). In this case, however, as is ob-

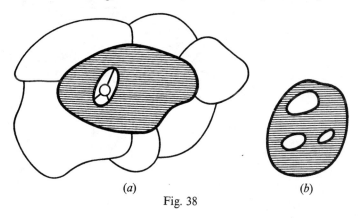

(a)   (b)

Fig. 38

vious from the figures, the region necessarily separates the remaining regions into at least two groups that are nowhere contiguous; hence the map is *not connected*.

If we limit ourselves to the consideration of *connected maps*, the boundaries of each region form a closed contour. This closed contour is similar to the perimeter of a polygon with straight sides, differing only in that the pieces between vertices are generally curved. (At individual points this contour can touch itself, as shown in Fig. 39.) By using this similarity between the regions of a connected map and ordinary polygons, we prove the celebrated theorem of Euler.[1]

Fig. 39

[1] Leonhard Euler (1707–1783), one of the greatest mathematicians, was a member of the St. Petersburg Academy of Science from 1727 to 1741 and from 1766 to the end of his life, and of the Berlin Academy of Sciences from 1741 to 1766.

EULER'S THEOREM. *If r is the number of regions, v the number of vertices, and b the number of boundaries of a connected map, then*

$$r + v = b + 2.$$

Proof. We first prove this theorem for maps of polygons whose sides are straight line segments. The sea in such a map is the outside of the polygon (in Fig. 40 the sea is shaded).

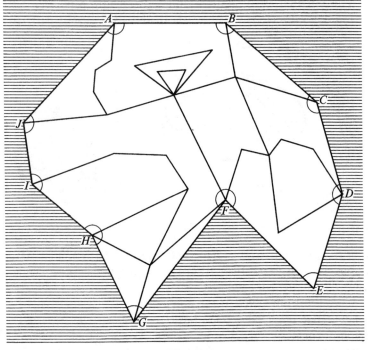

Fig. 40

Let us calculate the sum of the interior angles of all polygons in our map. The number of polygons is $r - 1$ (all regions except the sea). If a polygon has n sides, then, as is well known, the sum of its interior angles is $(n - 2) \cdot 180°$. Hence, the sum of the interior angles of all the polygons is

$$T = (n_1 - 2) \cdot 180° + (n_2 - 2) \cdot 180° + \cdots + (n_{r-1} - 2) \cdot 180°,$$

where $n_1, n_2, \ldots, n_{r-1}$ are the numbers of sides of our polygons.

This sum is equal to

$$T = [n_1 + n_2 + \cdots + n_{r-1} - 2(r-1)] \cdot 180°.$$

Let b_e be the number of outer boundaries (lying next to the sea), and b_i be the number of interior boundaries (not lying next to the sea). Then

$$b = b_i + b_e$$

(in Fig. 40, $b_i = 32$, $b_e = 10$). By Problem 19,

$$n_1 + n_2 + \cdots + n_{r-1} + b_e = 2b,$$
$$n_1 + n_2 + \cdots n_{r-1} = 2b - b_e.$$

Hence,

$$T = [2b - b_e - 2(r-1)] \cdot 180°.$$

We now calculate the same sum T in another way. Let v_i be the number of interior vertices and v_e be the number of vertices at the sea (in Fig. 40 $v_i = 19$, $v_e = 10$). Obviously,

$$v = v_e + v_i$$

and

$$b_e = v_e.$$

The sum of the interior angles of the polygons at each interior vertex is $2 \cdot 180°$. Hence, the sum of the interior angles of all the polygons, except those angles whose vertices are at the sea, is $2 \cdot 180° \, v_i$. To obtain T, we must add to this sum all the angles lying at the sea (they are marked in Fig. 40), in other words, the sum $(v_e - 2) \cdot 180°$ of all interior angles of the polygon $ABC \ldots J$. Thus,

$$\begin{aligned} T &= 2 \cdot 180° \, v_i + (v_e - 2) \cdot 180° \\ &= (v - v_e) \cdot 2 \cdot 180° + (v_e - 2) \cdot 180° \\ &= (2v - v_e - 2) \cdot 180°. \end{aligned}$$

If we set the two expressions for T equal to each other, we get

$$[2b - b_e - 2(r-1)] \cdot 180° = [2v - v_e - 2] \cdot 180°,$$
$$2b - b_e - 2(r-1) = 2v - v_e - 2.$$

Since $b_e = v_e$, it follows that

$$\begin{aligned} 2b - 2(r-1) &= 2v - 2, \\ b - r + 1 &= v - 1, \\ r + v &= b + 2. \end{aligned}$$

We now consider the general case, in which the regions can have curved boundaries also. As before, let v be the number of vertices, r the number of regions, and b the number of boundaries. On the boundaries, we introduce new, auxiliary vertices of multiplicity 2 (in Fig. 41 the vertices A, B, C, D, E, F, G, H, I). In this way, we

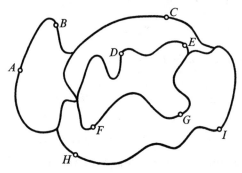

Fig. 41

obtain as many new boundaries (the pieces of old boundaries between the new vertices) as there are new vertices. Now we replace each curved boundary in the new map by a straight line segment with the same end points. If we place the auxiliary vertices sufficiently close together (we can introduce arbitrarily many of these vertices), these line segments will not cross. We obtain a new map (the dotted lines in Fig. 42), all of whose regions are polygons with straight

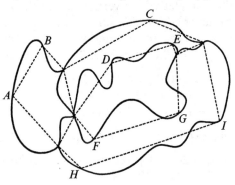

Fig. 42

sides. If the number of auxiliary vertices is equal to v' and the number of new boundaries is equal to b', then the total number of vertices in the new map is equal to $v + v'$ and the total number of boundaries is equal to $b + b'$. The number of regions, however, remains the same. As has just been proved, Euler's theorem holds for the new map; that is,

$$(v + v') + r = (b + b') + 2.$$

Since

$$v' = b',$$

we have

$$v + r = b + 2,$$

Problem 47. Let six points be given in a plane. Let every point be joined to four other points by curves that do not intersect. Prove that all the regions of the resulting map are triangles.

Problem 48. Let seven points be given in a plane. Prove that it is impossible to join them with nonintersecting curves in such a way that each point is joined with exactly four other points. (Use the result of Problem 4.)

Problem 49. Given three houses and three wells, prove that it is impossible to connect every house with every well by nonintersecting paths.

Problem 50. Given five points in a plane, prove that it is impossible to join every point with every other point by nonintersecting curves.

Problem 51. Prove that there cannot exist a map containing five regions with the property that every two neighbor each other.

Problem 51 suggests that four colors are sufficient for the proper coloring of any map. The proof of the five-color theorem is based on the result of this problem. Obviously, in the proof of this theorem, vertices of multiplicity two can be excluded from consideration. In Problems 52 to 54 we shall assume that the multiplicity of any vertex is at least three.

Problem 52. Prove that in any connected map in which any of the vertices has a multiplicity of at least three, there exists a region with fewer than six boundaries.

15. THE FIVE-COLOR THEOREM

Problem 53. Prove that *every map can be properly colored with six colors.*[1]

Problem 54. Prove the FIVE-COLOR THEOREM: *Every map can be properly colored with five colors.* (Use the result of Problem 51.)

We can solve the four-color problem for a special case with the methods used in the solution of the five-color problem. This is done in Problems 55 and 56. Exactly as in the problem on five colors, we assume that the map under consideration contains no vertex of multiplicity 2.

Problem 55. Prove that the inequality

$$b \leq 3r - 6$$

holds for connected maps in which the vertices have multiplicities of at least 3.

Problem 56. Prove that *any map with less than twelve regions can be properly colored with four colors.*

---

[1] It is assumed that the map has no boundaries that do not separate regions (see Definition in section 4).

Concluding Remarks

Problems—such as multicolor problems—about the properties of figures and solids which do not change under arbitrary deformations, in which the figures and solids are not torn or glued together, belong to a particular branch of mathematics called *topology*. Topology, which is one of the youngest branches of mathematics, developed into a distinct mathematical discipline about the turn of the last century. A leading role in the development of topology in the last thirty years has been played by the Soviet topological school, whose most prominent representatives are P. S. Urysohn (1898–1924), P. S. Alexandroff (1896–   ), and L. S. Pontryagin (1908–   ).

The reader can find accounts of questions having to do with Euler's theorem and multicolor problems in the following works whose presentations differ from this book's:

Rademacher, H., and Toeplitz, O. *Enjoyment of Mathematics.* Princeton, N. J.: Princeton University Press, 1957.

Hilbert, D., and Cohn-Vossen, S. *Geometry and the Imagination.* Translated by P. Nemenyi. New York: Chelsea Publishing Company, 1952.

In the latter book, problems of map-coloring on surfaces more complicated than the plane are solved. The reader will also find some elementary material on topology in it.

An elementary introduction to topology is presented in the book:

Arnold, B. H. *Intuitive Concepts in Elementary Topology.* Englewood Cliffs, N.J.: Prentice-Hall, Inc., 1962.

Selected research articles are listed in the Bibliography on page 68.

Appendix

COLORING A SPHERE WITH THREE COLORS

Suppose that a sphere[1] is divided into a certain number of domains; in other words, suppose a map is drawn on a sphere. We can ask if there exists a region containing antipodal points, that is, two diametrically opposite points of the sphere. If there are four regions, it is possible that no region exists that contains antipodal points (Fig. 43). If the number of regions exceeds four, then it is all

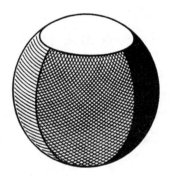

Fig. 43

the more true that such a region need not exist. On the other hand, if there are only two regions in the map, then at least one of them will contain both end points of a diameter of the sphere.[2] Indeed, it can happen that every such pair of points belongs to the boundary of both regions. An example is a sphere divided into two hemispheres.

---

[1] By a sphere we mean the *surface* of a ball.

[2] To prove this, it is sufficient to consider an arbitrary point A on the boundary between the two regions. Its antipodal point A' belongs to one of the regions. Since A belongs to that same region (a point on the boundary we consider as belonging to both adjacent regions), A and A' form the desired pair of points.

The question now arises, what about the case of three regions? We shall prove that here, as with two regions, there is always a region containing a pair of diametrically opposite points. Further, a far more general assertion can be proved.

THEOREM. *If a sphere is divided into an arbitrary number of regions (greater than two), and these regions are arranged in three groups in an arbitrary manner, then there exist two diametrically opposite points belonging to regions of the same group.*[1]

For clarity, we shall assume that the regions of each group are colored the same color: for example, the regions of the first group, blue; those of the second group, black; and those of the third group, red. A pair of diametrically opposite points belonging to regions of the same group we shall call *identically colored.*

Proof. We prove the theorem by contradiction. Assume that a map K drawn on a sphere is colored with three colors, and that no pair of diametrically opposite points lie in regions of the same color.

This assumption leads to a contradiction in the following way: We prove that from every three-color map with no identically colored pairs, a new map can be constructed that again contains no identically colored pairs, but has fewer regions than the first map. Thus, if we start off with the map K, whose existence we have assumed, we can construct a map K_1 with fewer regions and containing no identically colored pairs. Then from the map K_1 we can construct a map K_2 by the same method, and so on. We obtain an infinite sequence of maps

$$K, K_1, K_2, K_3, \ldots, K_m, \ldots,$$

none of which contains identically colored pairs. If we let n be the number of regions in the map K, n_i the number in K_i, then

$$n > n_1 > n_2 > n_3 > \cdots > n_m > \cdots.$$

Thus, we obtain an infinite decreasing sequence of positive integers, which is impossible. The exposition here of such a proof leads us to the following definition.

DEFINITION. *The* method of infinite descent *is a proof by contradiction, using an infinite decreasing sequence of positive integers.*

---

[1] This theorem was discovered (in a much more general form) by the Soviet mathematicians L. A. Lyusternik and L. G. Shnirelman.

Thus, still attempting to prove the theorem by contradiction, again let an arbitrary map K, containing no identically colored pairs, be given. A map K_1 must be constructed, again containing no identically colored pairs, but consisting of fewer regions than the map K.

Note that if the map K is not properly colored, that is, if it contains boundaries between regions of the same color, then we can immediately obtain the desired map K_1 by erasing these boundaries. Therefore, we shall consider henceforth only properly colored maps K.

We use the following lemma to prove our theorem.

LEMMA. *In a map with more than two regions, a point diametrically opposite a boundary point is an interior point; that is, it lies strictly inside some third region.*

Proof. Indeed, in a *properly colored* map of more than two regions, every boundary point belongs to at least two differently colored regions. For example, let a boundary point A belong to a red and a blue region; then its antipodal point A' can belong neither to a blue nor a red region. Therefore, it lies strictly within a black region (as according to the lemma).

Proof of the theorem, continued. Applying the lemma to the proof of the theorem, we let B be a point lying on the boundary of some region, colored red, say. We move from the point B along the boundaries of the map in such a way that red regions are always on our left (Fig. 44).[1] We continue until we return to a point C

Fig. 44

that we have already touched. We now consider the closed non-intersecting contour $CfgC$ (Fig. 44), which we denote by Γ. We shall draw the map on the sphere separately for the "northern" and "southern" hemispheres. We assume, for clarity, that the con-

---

[1] In Figures 44, 46, and 47 red is indicated by horizontal shading, blue by cross-hatching.

tour Γ lies wholly in the northern hemisphere (Fig. 45). For the general case, the proof is unchanged. Now we construct a contour

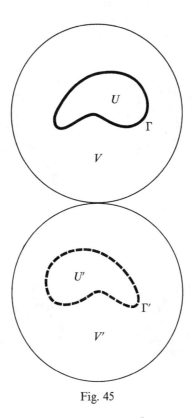

Fig. 45

Γ', consisting of the points diametrically opposite the points of the contour Γ (Fig. 45). By the lemma, the contour Γ' consists only of interior points, that is, it cuts none of the boundaries of the map. In particular, it does not cut the contour Γ, which is formed from boundaries.

The contour Γ divides the sphere into two parts, U and V (Fig. 45). In exactly the same way, Γ' divides the sphere into parts U' and V', where U' consists of points diametrically opposite those of U, and V' consists of points diametrically opposite those of V. Since Γ and Γ' do not intersect each other, the contour Γ' lies in one of the parts U or V. If in V, then the contour Γ lies in V'.

Since the contour Γ' does not cut any boundary of the map, it lies entirely inside some region. All points of the contour Γ belong to the red regions; hence, no point inside or on the contour Γ' can be red. To be specific, we can assume that this region is colored black (Fig. 46).[1] In this case, no point of the contour Γ can belong

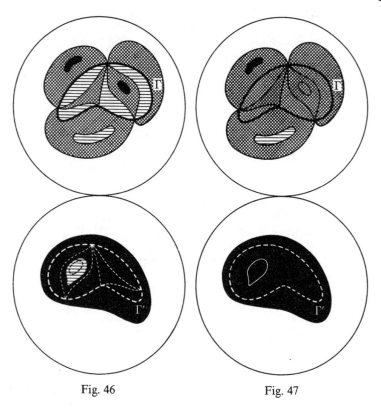

Fig. 46   Fig. 47

to a black region; in other words, all boundaries from which the contour Γ is formed are boundaries between blue and red regions. Since only red regions lie on one side of the contour Γ, then only blue regions lie on the other side (the map is properly colored). Suppose that red regions lie on the side U and blue ones on the side V, as in Figure 46. If now we change the color of all regions in U to blue, so that blue regions border on both sides of the contour Γ (Fig. 47), then at the same time all regions in U' remain black.

---

[1] In Figure 46 the dotted lines consist of points diametrically opposite to boundary points in the region U.

(Notice that no identically colored pairs occur in this coloring, since, first of all, the points in U have become blue and their antipodal points in U' black and, secondly, all other points, as well as their antipodal points, have not changed color.) Now let us erase the contour Γ and all boundaries lying inside U and U'. The number of regions is thereby decreased (since the contour Γ separated at least two regions from one another).

Thus, from an arbitrary map K containing no identically colored pairs, we have constructed a map K_1, which has fewer regions than the map K and, likewise, has no identically colored pairs. The method of infinite descent is used once more as in the first part of this proof, and our theorem is proved by contradiction.

Solutions to Problems

PROBLEM 1. We shall prove this theorem by mathematical induction. Assume the theorem already proved for n straight lines. We shall prove that it is then also valid for $n + 1$ straight lines. Let us consider a map K formed by $n + 1$ straight lines, and erase one of these lines, say the line l. We then have a map K^* formed by only n straight lines, which, according to the assumption, can be properly colored with two colors. We color this map, K^*, in black and white. We then replace the erased line l. It divides the plane into two parts, each of which is properly colored with two colors (Fig. 48a). We now leave the colors

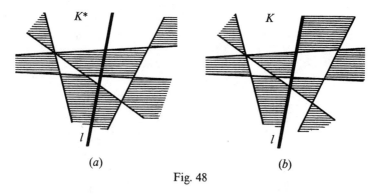

(a)   (b)

Fig. 48

of all the regions in one of the two parts unchanged, while in the other part we replace black by white and white by black (Fig. 48b). Each half-plane herewith remains properly colored. If two neighboring regions of the map K be in different half-planes, they bound one another along a segment of the straight line l. The new regions are formed by the dissection of some region of the map K^* by this straight line. In this case also, each two neighboring regions of the map K are colored in different colors. The theorem is valid for $n = 1$; according to what we have proved, it is valid for $1 + 1 = 2$ straight lines, for $2 + 1 = 3$ straight lines, etc.; thus, it is valid for any given number of straight lines.

PROBLEM 2. This problem can be solved by the method of mathematical induction, exactly as Problem 1 was solved. Instead of proceeding in this way, however (the reader can do it as a useful exercise), we shall carry out the following plan. For each of the regions into which the plane is divided, we count the number of circles within which it lies. (For the map in Fig. 5 the results of such counting are shown in Fig. 49.) We note that the numbers for

any two neighboring regions always differ by 1. In fact, if two neighboring regions A and B are separated by the arc of a circle C, then one of the regions lies inside C and the other outside C; and the regions A and B either both lie, simultaneously, inside or else both lie, simultaneously, outside every circle other than C. It is sufficient, therefore, to color all the even-numbered regions with one color, and all the odd-num-

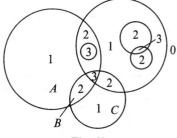

Fig. 49

bered regions with another color, in order to obtain a proper coloring of our map.

PROBLEM 3. In such a numbering of vertices, at the three vertices of each triangle the three digits, 0, 1, 2 will appear. We draw arrows along the sides of the triangles from 0 to 1, from 1 to 2, and from 2 to 0. In this way, each triangle receives a definite orientation. One can distinguish two types of triangles: triangles with clockwise orientation (Fig. 50) and triangles with

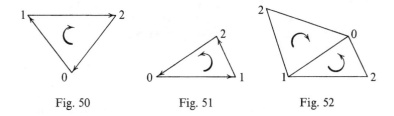

Fig. 50   Fig. 51   Fig. 52

counterclockwise orientation (Fig. 51). We color all triangles of the first type white and all those of the second type black. Since any two neighboring triangles always have opposite orientations (Fig. 52), we shall obtain a proper coloring.

PROBLEM 4. We assume the contrary, and color our map *properly* with white and black. Suppose that in this way we obtain k white regions having $n_1, n_2, \ldots, n_k$ boundaries, and l black regions having $n_1', n_2', \ldots, n_l'$ boundaries. Each boundary belongs to exactly one white and one black region; if b denotes the total number of boundaries, then

$$b = n_1 + n_2 + \cdots + n_k = n_1' + n_2' + \cdots + n_l'.$$

All but one of the numbers $n_1, n_2, \ldots, n_k, n_1', n_2', \ldots, n_l'$ is divisible by m.

It then follows, from this equality, that *all* the numbers of boundaries must be divisible by m, and this proves our premise by contradiction.[1]

PROBLEM 5. In Figure 53 the squares are numbered in the order in which the knight goes around them.

7	12	17	22	5
18	23	6	11	16
13	8	25	4	21
24	19	2	15	10
1	14	9	20	3

Fig. 53

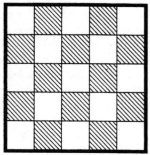

Fig. 54

PROBLEM 6. See Figures 53 and 54. We have obtained a proper coloring of the 25-square chessboard. In the same way, we obtain a proper coloring of any other board, such that the knight can make the rounds of all the squares touching each one only once, and coloring the even-numbered squares all one color. Suppose such a board is already properly colored with two colors, as may always be done.[2] Then the knight jumps from a square of one color to a square of the opposite color on each move. Therefore, if all the squares are numbered in the order in which the knight touches them, all squares with even numbers will be of one color, black, say, and all squares with odd numbers will be of the second color, white, say. To reverse the procedure, if, on an uncolored but numbered chessboard, we color all the regions with even numbers black, and all the regions with odd numbers white, we simply revert to the previous coloring, which is a *proper* coloring.

[1] Suppose, to be specific, it is n_1 that is not divisible by m. Since $b = n_1' + n_2' + \cdots + n_i'$ and $n_1', \ldots, n_i'$ are all divisible by m, b must be divisible by m. But then, since $n_1 = b - (n_2 + \cdots + n_k)$, and $n_2, \ldots, n_k$ are all divisible by m, n_1 must be divisible by m. Here is a contradiction of the supposition that n_1 is not divisible by m; therefore, the theorem is proved.

[2] While it is true that every chessboard can be properly colored with two colors in the customary checkerboard design, it is not true of every chessboard that a knight can make the rounds of all the squares so as to touch each square only once. A 2×2 chessboard is obviously such an exception. Find another one.

PROBLEM 7. No, it is impossible. Let us color the board properly with two colors, then number the squares in the order in which the knight made the rounds of them. The first square has the number 1; the last, the number 49. They ought, therefore, to have the same color, as do all the squares with odd numbers (see the solution to the preceding problem). But two neighboring squares in a proper coloring are colored differently.

PROBLEM 8. Let us color our board properly with two colors, and in such a way that the square S is colored white (Fig. 55). If a knight could tour the board starting from the square S, then, numbering the squares in the order of passage, we would have assigned the odd numbers 1, 3, ..., 49 to the white squares (compare with the solution of Problem 6). The number of white squares would thus be 25. But in all there are actually only 24 white squares, as Figure 55 clearly indicates.

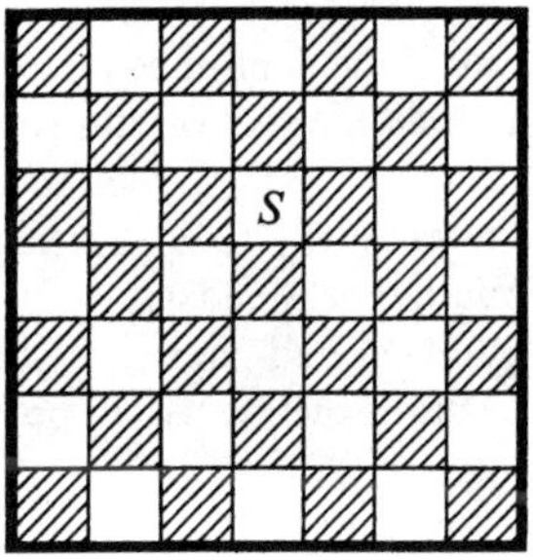

Fig. 55

PROBLEM 9. With each move, the knight changes the color of his square. Therefore, when he reaches a region of the same color as the one from which he started (in the given case he has reached the actual square from which he started), he has made an even number of moves.

PROBLEM 10. Each of the rook's moves can be replaced by a series of simple moves, each of which is a move from a square to a neighboring one. With each such simple move, the rook changes the color of its square. In a tour of the whole board of 64 squares, it will make 63 simple moves, an odd number. Hence, the last square it reaches must have a color different from that of the square where it started, while the squares A and B have the same color.

PROBLEM 11. Impossible. A domino consists of two halves, each half bearing one of the seven numerals 0, 1, 2, 3, 4, 5, 6. There are as many dominoes as there are possible pairs of these numerals. Each numeral thus occurs eight times, six times with other numbers, and twice more on one domino, a double, on which the number is combined with itself. In the chain of dominoes, two dominoes are joined together so that the two touching halves bear identical numerals. Hence, each numeral occurs an even number of times, say n, in the interior of the chain, and $8 - n$ times on the ends, also an even number of times. Consequently, every number either appears at neither end (and zero is an even number), or at both ends at the same time.

PROBLEM 12. Let N be the total number of human beings who ever lived on earth, and let m be the number of handshakes they have exchanged. Number all human beings, and denote by n_k the number of times the kth human being has shaken hands. If the kth human being has shaken hands with the lth, then this handshake is included once in the number n_k and again in the number n_l. Hence, in the sum

$$S = n_1 + n_2 + n_3 + \cdots + n_N \quad (N \text{ summands}),$$

each handshake is counted twice and

$$S = 2m.$$

But, if the sum of any group of numbers is even, as here, then there must be an *even* number of odd summands.

PROBLEM 13. If all participants in the meeting shook hands with an odd number of friends, it would follow that each of the 225 persons (225 is an odd number) had shaken hands an odd number of times. This contradicts the result of the preceding problem (with $N = 225$).

PROBLEM 14. If this were possible, then from each point there would emanate three curves. If we multiply by 3 the number of points, here 5, we would be counting each curve twice (since each curve has two end points) and would thus obtain double the number of curves, an even number. But $5 \cdot 3 = 15$, an odd number.

PROBLEM 15. One possible solution for (a) is exhibited in Figure 56, and a solution for (b) is drawn in Figure 57.

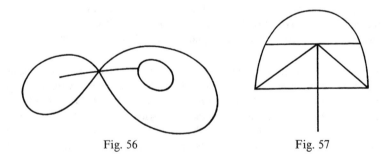

Fig. 56   Fig. 57

PROBLEM 16. In the sum $k_1 + k_2 + \cdots + k_v$ each boundary is counted twice, since it is counted in the multiplicity of each of its end points. (If the end points of a boundary coincide, the boundary is counted twice in determining the multiplicity of that end point.)

PROBLEM 17. The statement follows from Problem 16. Indeed, if the sum of some numbers is even, then the number of odd summands occurring must be even.

PROBLEM 18. Possible solutions for (*a*), (*b*), (*c*) are shown in Figures 58, 59, 60, respectively.

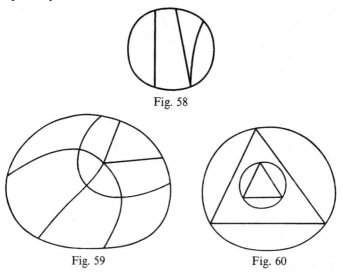

Fig. 58

Fig. 59   Fig. 60

PROBLEM 19. Every boundary is counted twice in the sum $n_1 + n_2 + \cdots + n_r$, since it belongs to two neighboring regions.

PROBLEM 20. This follows from Problem 19 in exactly the same way that Problem 17 follows from Problem 16.

PROBLEM 21. See Figures 61 and 62.

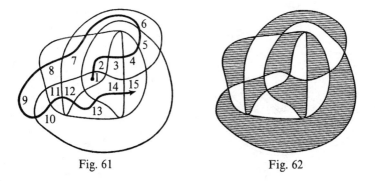

Fig. 61   Fig. 62

PROBLEM 22. Let us properly color our map with two colors (Fig. 63). If we number the regions in the order in which they are toured, all regions with

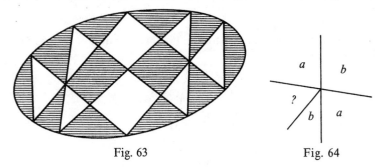

Fig. 63   Fig. 64

even numbers will be of one color, and all with odd numbers of the second color (see the solution of Problem 6). In all, there are 21 regions, of which 11 must have odd numbers (1, 3, ..., 21), and 10 have even numbers (2, 4, ..., 20). At the same time, however, there are 12 black and only 9 white regions.

PROBLEM 23. If any vertex has an odd multiplicity, then even the regions surrounding it cannot be properly colored with two colors (Fig. 64).

PROBLEM 24. Let us consider the path of our rook (the dotted line in Fig. 65). We now erase part of the map lying outside the path, and adjoin the path itself to the map. We obtain a new map (Fig. 66). All interior vertices of this map have, by assumption, an even multiplicity, while the outer vertices, formed by the intersection of the boundaries of the old map with the path of the rook, all have multiplicity 3. There will be an even number of vertices having odd multiplicities (Problem 17); therefore, the rook must have crossed

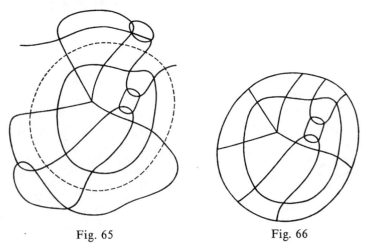

Fig. 65   Fig. 66

over an even number of boundaries or, what amounts to the same thing, it has made an even number of moves.

PROBLEM 25. The case of the rook starting out from S_0 and visiting each region not more than once has already been considered in Problem 24. Suppose now that the rook has visited some region S_1 twice, that is, it has crossed its own path in this region (Fig. 67). Let the number of moves in the segment S_0aS_1 be p, the number in S_1bS_1 be q, and the number in S_1cS_0 be r. We have to prove that $p + q + r$ is even. By problem 24, q is even. The path $S_0aS_1cS_0$ also satisfies the conditions of Problem 24; hence $p + r$ is likewise even. From this it follows that $p + q + r$ is even. If, on the other hand, the rook makes two loops instead of one (Fig. 68), then, by erasing one of them, S_1bS_1, for example, we obtain the case just considered. An even number of moves, again by Problem 24, is made in going around the loop S_1bS_1; hence, an even number of moves will be needed for the entire path with S_1bS_1 included. In this way, we can prove our statement for an arbitrary number of loops.

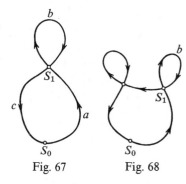

Fig. 67   Fig. 68

PROBLEM 26. If the second path is traversed in the opposite direction, then we go from S_1 to S_0 in q moves. Therefore, if we go from S_0 to S_1 along the first path, and then from S_1 to S_0 along the second path in the reverse direction, we return to S_0 in $p + q$ moves. By Problem 25, $p + q$ is an even number, whence p and q are either both even or both odd.

PROBLEM 27. *First method.* Let us start out with the rook from some region S_0, and traverse all the regions of the map, one after another. We number them in the order in which the rook has visited them. It may happen that some regions are visited twice, or even oftener. Any such region is assigned, not one number, but two or more numbers. By Problem 26, however, the numbers assigned to any one region will be either all even or all odd. On the other hand, neighboring regions must have numbers that differ by an odd number. In fact, if region S_1 can be reached from S_0 in p moves and if S_2 borders on S_1, then S_2 can be reached from S_0 in $p + 1$ moves. This means that every path leading from S_0 to S_2 consists of a number of moves that is even if $p + 1$ is even, and odd if $p + 1$ is odd; therefore, this number is even if p is odd, and odd if p is even. If, after the rook has visited all regions of the map, we color with one color all regions with even numbers, and use a second color for all regions with odd numbers, then we obtain a proper coloring.

Second method. We shall travel over the boundaries and vertices of our map. In the process, we always leave from a vertex along a boundary different from that on which we arrived. Since there are no vertices of multiplicity 1 on our map, having arrived at an arbitrary vertex, we can always go out from it. We continue along in this fashion until we reach, for the first time, a vertex already met before, say the vertex *A*. The part of our path traversed in going out from the vertex *A* and returning to it forms a closed contour that does not intersect itself; we erase this from our map. In consequence, the multiplicity of each vertex either is not altered (if the contour does not pass through it) or is decreased by 2 (if the contour does pass through it). Some vertices of multiplicity 2 may thereby vanish, but, as before, the multiplicity of any vertex remains even. On the newly obtained map, we again single out some closed contour which consists of boundaries and does not intersect itself, and we likewise erase it. We continue the process until the whole map has been erased. One can, consequently, obtain a map all of whose vertices have even multiplicities by laying on top of one another these closed contours, each of which divides the plane into two parts. This map is entirely analogous to the map in Figure 5, which is obtained by the superposition of circles. Exactly as in Problem 2, we can prove that such a map can be properly colored with two colors.

Remark. With the help of analogous considerations, it is not difficult to show the following. We can draw a map, all vertices of which have even multiplicities, with one stroke of the pen, that is, without lifting the pen from the paper or tracing over any boundary twice. In general, we can draw a network of curves in one stroke of the pen in only two cases, either if all of its vertices have even multiplicities, or if exactly two of the vertices have odd multiplicities. (In the latter case we must, however, begin the stroke in one of these two vertices and end in the other.)

PROBLEM 28. To solve this problem, we can either use induction (see Problem 1) or apply the method used in the solution of Problem 2. We give here an outline of the proof by the second method. We select any one of our figures (a circle with a chord). It divides the plane into three parts. We number all regions of the map that lie in one of these parts, 0; all that lie in the second part, 1; and all that lie in the third part, 2. We do this for all the figures. Each region will then have *n* numbers. We add them and find the remainder after division by 3. The regions for which this remainder is 0, we color white; those for which the remainder is 1 or 2, we color red or black, respectively. It can now be shown, exactly as in Problem 2, that the resulting coloring is proper.

PROBLEM 29. Let a side of the board be made up of m hexagons. We remove the $6(m - 1)$ regions lying on the six sides of the board. We then obtain a board of the same form, but whose sides consist of only $m - 1$ hexagons.

Hence, if we denote the number of regions of the board with m sides by S_m, then

$$S_m = 6(m - 1) + S_{m-1}.$$

In exactly the same way,

$$S_{m-1} = 6(m - 2) + S_{m-2},$$
$$S_{m-2} = 6(m - 3) + S_{m-3},$$
$$\cdots\cdots\cdots\cdots\cdots\cdots\cdots$$
$$S_2 = 6 \cdot 1 + S_1,$$
$$S_1 = 1.$$

From this, it follows that

$$S_m = 6(m - 1) + 6(m - 2) + \cdots + 6 \cdot 1 + 1$$
$$= 1 + 6[1 + 2 + \cdots + (m - 2) + (m - 1)].$$

But[1]

$$1 + 2 + \cdots + (m - 2) + (m - 1) = \frac{(m - 1)m}{2}.$$

Therefore,

$$S_m = 1 + 6\frac{(m - 1)m}{2} = 3m^2 - 3m + 1.$$

PROBLEM 30. See Figure 69.

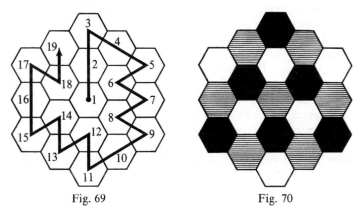

Fig. 69   Fig. 70

PROBLEM 31. We obtain a proper coloring in three colors for a board each of whose sides contains three hexagons (see Fig. 69 and 70).[2] We obtain the

[1] See, for example, I. S. Sominskii, *The Method of Mathematical Induction* (Boston: D. C. Heath and Company, 1963).

[2] In Figure 70 red is indicated by horizontal shading.

same result on an arbitrary board. In fact, assume that we have a board already properly colored with three colors.[1] In this case, the camel goes from red regions to white, from white to black, and from black to red (compare with Fig. 19 and 20). If the camel begins its path in the central red region, and we write down the sequence of colors of the regions through which it moves, we obtain the sequence (r = red, w = white, b = black):

$$rwb \quad rwb \quad rwb \quad rwb \quad rwb \quad \ldots .$$

Hence, regardless of the path the camel takes (assuming, however, that the path begins in a red region and that the second region is colored white), all regions whose number is divisible by 3 will be black, and all regions whose number gives a remainder of 1 or 2 when divided by 3 will be red or white, respectively.

Thus, the coloring described in the formulation of Problem 31 coincides, in fact, with the initial proper coloring of the hexagonal board (compare with Problem 6).

PROBLEM 32. For example, if the camel begins its path in a black region, then the sequence of the colors of the regions through which it passes will be (see the previous problem)

$$brw \quad brw \quad brw \quad b.. \quad \ldots .$$

The period of this sequence consists of three terms; therefore, the camel can reach a region of the same color as the region from which it started (in particular, it can reach the actual region from which it started) if and only if the number of moves it has made is divisible by 3.

PROBLEM 33. To tour a board with 19 regions (see Fig. 70), the camel must make 18 moves, that is, a number of moves which is a multiple of 3, and arrive at a region of the same color as that from which it started. Hence, it will have passed through more regions of this color than regions of the other two colors. However, there are no more regions colored like the corner region than there are regions of each of the other two colors (see Fig. 70).[2]

PROBLEM 34. To tour a board of $3m^2 - 3m + 1$ regions (see Problem 29), a camel must make $3m^2 - 3m = 3(m^2 - m)$ moves, and since the number of

---

[1] It is easy to see that a hexagonal board can always be properly colored with three colors. For instance, it is possible to apply the coloring of Figure 70 to a small section of the board and then to extend it in the obvious way to cover the entire board.

[2] On the contrary, there are more red regions than white or black. Hence Problem 30 does have a solution.

moves is a multiple of 3, on the last move it will reach a region of the same color as the one from which it started. But the regions neighboring this region all have a different color from it. Hence, the camel cannot finish its tour in a region which adjoins the region from which it started.

PROBLEM 35. The last move of the camel leads to a region of the same color as that from which it started. Hence, there would have to be more regions of that color than of either of the other two colors. However, on our board there are 21 black regions, 21 white regions, and 19 red regions (see Fig. 20).

PROBLEM 36. See Figures 71 and 72.[1] We obtain a proper coloring.

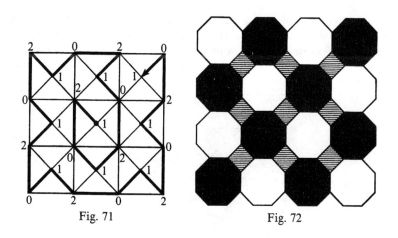

Fig. 71   Fig. 72

PROBLEM 37. Our board consists of 25 regions. To tour it in the specified way, 24 moves are necessary. On every third move, the camel reaches a region of the same color as the region from which it started (Fig. 72). Thus, on the last move, the camel must reach a region of the same color as the starting region, and there would have to be more regions of this color than of either of the other colors. But while the board has 9 of the square red regions, it has only 8 each of the octagonal black and white regions.

[1] In Figure 72 red is indicated by horizontal shading.

PROBLEM 38. If we pay no attention to the directions of the arrows, we can reach any vertex from any other vertex along the sides of the triangles. We now alter our path in such a way that it always conforms to the directions of the arrows. The path can be corrected in the following way: If we have gone along some side in the direction of its arrow, we leave that part of the path unchanged. If we have gone along a side in the direction opposite its arrow, we choose a path along the two other sides of one of the neighboring triangles (Fig. 73); the path from A to B goes against the direction of the arrow and we can therefore substitute either the path ACB or the path ADB.

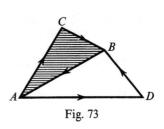

Fig. 73

PROBLEM 39. First of all, it is clear that it is sufficient to prove this theorem for the case where the camel never crosses its own path. Loops can be handled exactly as in the solution of Problem 25. Thus, let the camel have touched each vertex of its path only once and returned to its starting point. Now we erase everything outside of the region bounded by the path of the camel (Fig. 74). The map that results can be properly colored with two colors.

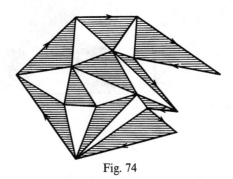

Fig. 74

Then all triangles lying to the right of the path of the camel are colored black, and those to the left are colored white. Hence, all triangles bordering the outside of the region have the same color, and we obtain a proper coloring if we color the exterior region the opposite color.

Since all regions except one in the map obtained are triangles, the number of boundaries of the exterior region, that is, the number of boundaries that lie on the path of the camel, is divisible by 3, as shown in Problem 4, q.e.d.

PROBLEM 40. By Problem 38, we can construct an auxiliary path from B to A. Suppose the camel needs r moves to traverse this path (Fig. 75). By Problem 39, $p + r$ and $q + r$ are both divisible by 3; hence,

$$(p + r) - (q + r) = p - q$$

is also divisible by 3.

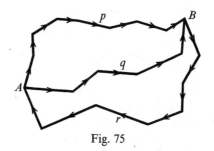

Fig. 75

PROBLEM 41. Let A and B be arbitrary vertices. In general, a camel can go from A to B by many different paths, where each path requires a certain number of moves. However, by Problem 40, the number of moves in each case must have the same remainder on being divided by 3. We take this remainder as the number of the vertex B. We do the same thing with all the other vertices. They will then be numbered with the digits 0, 1, and 2. Let us prove that neighboring vertices have different numbers. Let B_1 and B_2 be two such vertices. They are connected by an arrow, which points, say, from B_1 to B_2. Suppose the camel goes from A to B_1 in p moves, and then makes one more move in the direction of the arrow B_1B_2, reaching B_2. We have now gone from A to B_2 in $p + 1$ moves. But the numbers p and $p + 1$ give different remainders after division by 3.

PROBLEM 42. We construct the dual map K^* of the normal map K. According to condition 4 on page 20, the map K^* is made up of triangles. By the theorem on page 22, the map K can be properly colored with three colors if and only if the vertices of K^* can be properly numbered with three digits. But K^*, which is made up of triangles, has vertices that can be properly numbered with three digits if and only if the triangles can be properly colored with two colors (see pages 18–19). In other words, all the vertices of the map K^* must have even multiplicity. The last of these conditions is, however, equivalent to the condition that each region of the map K has an even number of boundaries (dual map condition 4). Thus, our assertion is proved.

Strictly speaking, this proof holds only for connected maps; only in this case can we pass to the dual map without any complications. The theorem is also true, however, for disconnected maps. Let us prove it for the case when the map falls into two completely separate parts. Let K be a disconnected normal map whose regions all have an even number of boundaries, and let S be the region that separates the regions of the map K into two groups, nowhere contiguous (Fig. 76). We consider separately the two maps K' and

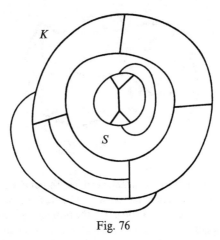

Fig. 76

K'' (Fig. 77a and b) into which the map K is divided. Each of these two maps is connected. All regions of the maps K' and K'', with the possible exceptions of S' and S'', have, by assumption, an even number of boundaries. By Problem 20, however, there cannot exist just *one* region with an odd num-

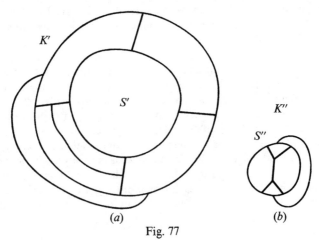

(a)   (b)

Fig. 77

ber of boundaries in the map K' (or in the map K''). S' and S'' must also have an even number of boundaries. By the theorem just proved, the maps K' and K'' can be properly colored with three colors. One must make sure that S' and S'' are colored with the same color. Then the map K can also be properly colored with three colors. The theorem can be proved in a similar way for the cases when the map is separated into three, four, or more parts.

PROBLEM 43. With each boundary, we associate the sum of the numbers of the two regions lying on each side of it. We obtain one of the three numbers $(0,1)$, $(1,0)$, or $(1,1)$ since the sum of two numbers is equal to $(0,0)$ only if the two numbers are equal. This numbering is proper, since, if a, b, and c are the numbers of the regions that meet at any vertex, the boundaries meeting at this vertex have the numbers $a + b$, $a + c$, and $b + c$. But $a + b \neq a + c$, since $b \neq c$. In exactly the same way, $a + b \neq b + c$ and $a + c \neq b + c$. (We have denoted three arbitrary choices of the number-pairs $(0,0)$, $(0,1)$, $(1,0)$, and $(1,1)$ in shortened notation by a, b, and c.)

PROBLEM 44. It is sufficient to consider the case when the rook visits each region no more than once. In the general case, we can proceed exactly as in Problem 25. (If the rook makes loops and the sum of the numbers in each loop is equal to $(0,0)$, then the sum of the numbers on the entire path is also equal to $(0,0)$.)

Let the vertices $A_1, A_2, \ldots, A_v$ lie inside the region bounded by the path of the rook (in Figure 78 that path is represented by a dotted line). On add-

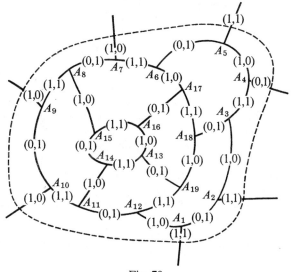

Fig. 78

ing the numbers of all the boundaries coming together at each vertex, we obtain (0,0). We write these equalities for each vertex, $A_1, A_2, \ldots, A_v$, and add. We obtain (0,0) on the right side, and on the left side, the sum of the numbers of all the boundaries one or both of whose ends lie inside the region through which the rook passes. The boundaries both of whose ends lie inside the region are counted twice, and the boundaries only one of whose ends lies inside the region—once. Let the sum of the numbers of the former be equal to x and of the latter be equal to y; then $2x + y = (0,0)$. But $2x = (0,0)$ for any number x; hence, $y = (0,0)$.

PROBLEM 45. *First method.* Fix any region S and let it have the number (0,0). Let the rook start out from this region and tour all of the regions of the map in such a way that each region is visited at least once. We add the numbers of all the boundaries that the rook has crossed on its path from S to a given region Q. Let this number be a, and let a be the number of the region Q.

First, we verify that the numbering of the regions is independent of the path of the rook. Indeed, suppose that we have gone from S to some region Q along two paths (Fig. 79). Let the sum of the numbers of the boundaries

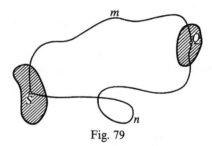

Fig. 79

on the path SmQ be a and on the path SnQ be b. By Problem 44, the sum of the numbers of the boundaries on the path $SmQnS$ is (0,0); consequently, $a + b = (0,0)$. But the sum of two numbers is equal to (0,0) only if these numbers are equal; hence, $a = b$, and Q receives the definite number a.

Let us now show that the numbering obtained is proper. Let P and Q be two neighboring regions separated by a boundary with some number c, and let the sum of the numbers of the boundaries on the path from S to P be a, so that P has the number a. We go from P to Q, crossing over the boundary with number c. Then Q has the number $a + c$, and $a + c$ is always different from a.[1]

[1] $a + c \neq a$ provided we number the boundaries with the pairs (0,1), (1,0), and (1,1).

Second method. We number the boundaries with the digits 1, 2, and 3. We start out from some vertex A along a boundary with number 1. We continue to move along the vertices and boundaries of the map in such a way that we go along a boundary with the number 2 after going along a boundary with the number 1, and we go along a boundary with the number 1 after going along a boundary with the number 2. We stop as soon as we reach a vertex B at which we have already been.

The boundary by which we arrive at B has either the number 1 or the number 2. Since the numbering is proper, the vertex B must coincide with A. (If the vertex B did not coincide with A, the numbering of the boundaries meeting at B would not be proper (Fig. 80).) Thus, our path, consisting of boundaries with the numbers 1 and 2, proves to be a closed contour that does not intersect itself. Let us choose an arbitrary boundary that has one of the numbers 1 or 2 and does not belong to the above contour, and begin the same process on it; we obtain a new contour. We continue this process until all the boundaries with numbers 1 and 2 are exhausted.

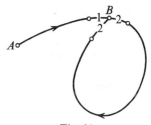

Fig. 80

Thus, the boundaries with the numbers 1 and 2 form a system of closed contours that do not intersect themselves. These contours partition the map into the domains $M_1, M_2, \ldots, M_p$; each domain contains several regions. If all of the boundaries with the number 3 are erased, a map remains whose boundaries are our contours and whose regions are the domains $M_1, M_2, \ldots, M_p$. The proof that $M_1, M_2, \ldots, M_p$ can be properly colored with two colors a and b can be taken verbatim from Problem 2. Let us consider an analogous system of contours consisting of the boundaries with numbers 1 and 3. These partition the plane into the domains $N_1, N_2, \ldots, N_r$, which we can color properly with the colors c and d. Each region S of the original map is contained in one of the domains $M_1, M_2, \ldots, M_p$ which has color a or b, and also in one of the domains $N_1, N_2, \ldots, N_r$ having color c or d. Hence, one of the pairs (a,c), (a,d), (b,c), (b,d) is associated with each region. We can easily verify that the numbering of the regions by such pairs is proper.

PROBLEM 46. According to Problem 45, it is sufficient to show that the boundaries of this map can be properly numbered with three numbers. Let a camel move on the boundaries of the map, going from any boundary to any neighboring boundary. We start out from an arbitrary boundary, to which we give the number 1, and number from there on as follows: After we

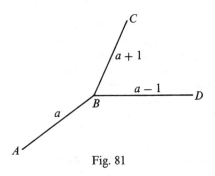

Fig. 81

have gone along any boundary AB, which has the number a (Fig. 81), we reach the vertex B. We can continue from there, either to the left (on BC) or to the right (on BD). If we go to the left, we increase the number by 1; that is, we give the number $a + 1$ to the boundary BC. On the other hand, if we turn right, we subtract 1 from the number a; that is, we give the number $a - 1$ to the boundary BD. In general, each boundary thereby obtains not one, but several integers (we remark that in this connection negative integers can also appear). Let us show that all numbers belonging to any given boundary give the same remainder after division by 3. (If we achieve this, we obtain a proper numbering by associating the corresponding remainder with each boundary.) The proof depends on the following

LEMMA. *A camel starts out from a boundary with the number p and returns again. The boundary thereby obtains a number q such that q − p is divisible by 3.*

Proof. Let the path of the camel contain b' boundaries, where it goes to the left b_1 times and to the right b_2 times ($b' = b_1 + b_2$). Obviously, $q = p + b_1 - b_2$; we must, therefore, show that $b_1 - b_2$ is divisible by 3. We know that it is sufficient to prove it for the case when the camel traverses no boundary more than once. The path of the camel is shown in Figure 82

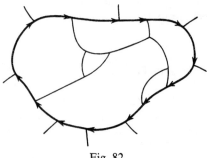

Fig. 82

($b' = 13$). All boundaries that end at vertices in the path of the camel, but that do not themselves belong to this path, fall into two kinds, those lying on the right with respect to the direction of motion of the camel (there are b_1 of them[1]) and those lying on the left (there are b_2 of them). If the camel goes around its path clockwise (Fig. 82), the boundaries of the first type lie inside the domain G which is bounded by the path of the camel, and the boundaries of the second type lie outside of this domain (in Fig. 82 $b_1 = 5$ and $b_2 = 8$). Thus, it must be proved that the difference between the number of boundaries diverging from the path of the camel toward the interior and the number of boundaries diverging from the path of the camel toward the exterior is divisible by 3.

Let v be the number of vertices that lie either inside the domain G or on the actual path of the camel; let b_i be the number of boundaries that lie inside the domain G, and $n_1, n_2, \ldots, n_r$ be the numbers associated with the boundaries of the regions that lie in G. The following relations hold:

(1) $$3v = 2(b_i + b') + b_2,$$
(2) $$n_1 + n_2 + \cdots + n_r = 2b_i + b'.$$

If we subtract (2) from (1), we obtain

$$b' + b_2 = 3v - (n_1 + n_2 + \cdots + n_r).$$

Since, by assumption, each of the numbers $n_1, n_2, \ldots, n_r$ is divisible by 3, $b' + b_2$ is also divisible by 3. But $b' = b_1 + b_2$; hence, $b_1 + 2b_2$ is divisible by 3, as well as

$$b_1 + 2b_2 - 3b_2 = b_1 - b_2.$$

We limit ourselves here to the proof of the lemma and leave it to the reader to complete the solution of the problem independently. (The outline of the proof is the same as in Problems 39 through 41.)

PROBLEM 47. On the resulting map, each vertex has multiplicity 4. By Problem 16,

$$4 \cdot 6 = k_1 + k_2 + \cdots + k_6 = 2b,$$

from which it follows that $b = 12$.

The map is connected. If it were not connected, at least one of the parts into which it was separated would have fewer than four vertices, since the total number of vertices is equal to six. Since this part would be completely isolated, each of the vertices could be connected only to a vertex of the same part. This contradicts the condition that each vertex be connected with exactly four other vertices.

---

[1] A boundary, both of whose end points lie on the path of the camel, will be counted twice.

Since the map is connected, we can apply Euler's theorem:

$$r = b + 2 - v = 8.$$

By Problem 19,

$$n_1 + n_2 + \cdots + n_8 = 2b = 24.$$

There are no 1-gons or 2-gons on the map (there are no vertices connected with themselves, and no two vertices are connected by two different boundaries). Hence, none of the numbers $n_1, n_2, \ldots, n_8$ is less than three. If one of them were greater than three, their sum would be greater than $3 \cdot 8 = 24$, which is not the case. Hence, $n_1 = n_2 = \cdots = n_8 = 3$. (The corresponding map is shown in Fig. 83.)

Fig. 83

PROBLEM 48. We assume the contrary to be true and draw the specified lines. We obtain a map all of whose vertices have multiplicity four. On the basis of Problem 16,

$$b = \frac{k_1 + k_2 + \cdots + k_7}{2} = \frac{4 \cdot 7}{2} = 14.$$

By Euler's theorem,

$$r = b + 2 - v = 9$$

(it can be proved, exactly as in the previous problem, that the map is connected). By Problem 19,

$$(1) \qquad n_1 + n_2 + \cdots + n_9 = 28.$$

Since the number of boundaries of each region is not less than three (see the solution of the previous problem), it follows from formula (1) that there are eight triangles and one quadrilateral. By Problem 4, our map cannot, therefore, be properly colored with two colors. But the multiplicity of each vertex is four, that is, even; this contradicts Problem 27.

PROBLEM 49. Let us assume these paths have been drawn. Then we obtain a map for which the vertices are houses and wells, and the boundaries are the paths. The number of vertices is six, and the multiplicity of each is three. The number of boundaries is $b = \frac{3 \cdot 6}{2} = 9$ (Problem 16), and the number of regions is $r = b + 2 - v = 5$. Since no path joins a house with a house or a well with a well, if we assign the digit 0 to every house and the digit 1 to every well, we obtain a proper numbering of our map. From this, it follows that each region has an even number of boundaries (the dual to Problem 23).

There are no 2-gons; hence, no region has fewer than four boundaries. By Problem 19,

$$18 = 2b = n_1 + n_2 + n_3 + n_4 + n_5 \geq 4 \cdot 5 = 20,$$

which is a contradiction.

PROBLEM 50. Let us assume the contrary. Then we have a map that has five vertices, the boundaries being the lines drawn between them. The multiplicity of each vertex is four. We have $b = \dfrac{4 \cdot 5}{2} = 10$ (Problem 16). Hence, by Euler's theorem, $r = b + 2 - v = 7$. By Problem 19,

$$n_1 + n_2 + \cdots + n_7 = 2b = 20.$$

On the other hand, since $n_1, n_2, \ldots, n_7$ are all no less than three,

$$n_1 + n_2 + \cdots + n_7 \geq 3 \cdot 7 = 21.$$

But this is a contradiction.

Remark. Although in the plane it is impossible to join five points pairwise with nonintersecting curves, this problem is easily solved in space (Fig. 84).

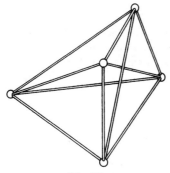 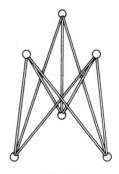

Fig. 84   Fig. 85

Analogously, solutions for Problem 49 (Fig. 85) and Problem 48 as well can be constructed in space. In general, any figure consisting of n points joined by lines that intersect in the plane can be constructed in space without intersections. (A net is formed with nonintersecting cords, where the n points to be joined are the vertices of the net.) Difficulties arise in the solution of the converse problem: Which networks of curves in space can be portrayed in the plane without intersections? Figures 84 and 85 show examples of networks in space whose representation in the plane is impossible without intersections. It can be proved that all networks that cannot be represented in the plane without intersections are similar to these; that is, each such network contains (in the sense mentioned) a configuration similar to the one in Problem 49 or the one in Problem 50.

PROBLEM 51. If such regions existed, we could choose a capital in each and connect the capitals of neighboring regions with railroad lines (in other words, pass to the dual map). We would then obtain a configuration that was shown to be impossible in the preceding problem. This problem, like the previous one, can be solved by direct calculation.

PROBLEM 52. We assume the contrary to be true. Then each of the numbers $n_1, n_2, \ldots, n_r$ is no less than six and

(1)
$$6r \leq n_1 + n_2 + \cdots + n_r = 2b, \quad 3r \leq b.$$

On the other hand, the multiplicity of any vertex is no less than three; hence (Problem 16),

(2)
$$3v \leq 2b.$$

By adding the inequalities (1) and (2), we obtain

$$3(v + r) \leq 3b,$$
$$v + r \leq b,$$

which is impossible, because $v + r = b + 2$.

PROBLEM 53. The proof is by induction. The theorem obviously holds for a map with at most six regions. Assume it has been proved for a map with n regions. We shall show that it then also holds for a map of $n + 1$ regions.

By the previous problem, there exists a region S with less than six boundaries (Fig. 86a). We now assume that the plane on which our map is drawn

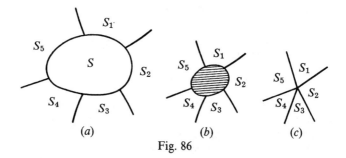

Fig. 86

consists of a rubber film, and we carry out the following operation. We cut out the region S with scissors, making a hole (Fig. 86b), and contract the edges of the hole until it is completely closed (Fig. 86c). We then obtain a map consisting of n regions; by assumption, this map can be properly colored with six colors. If we now again insert the region S, we can color it with a color different from the colors of all the neighboring regions, since it is a neighbor of no more than five regions. Thus, the theorem is proved.

The proof has been carried through only for connected maps (we used Problem 52). However, it can be adapted directly to disconnected maps. We shall demonstrate how this is done by an example. A disconnected map with the region S separating its regions is drawn in Figure 87. The boundaries of

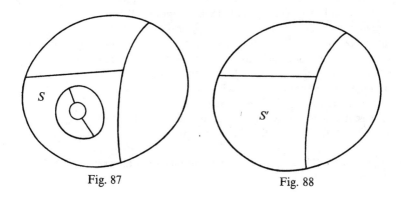

Fig. 87   Fig. 88

this map fall into two connected parts, which are drawn separately in Figures 88 and 89. The maps in Figures 88 and 89 are connected; it has been proved

Fig. 89

that they can be properly colored with six colors. We arrange it so that S' and S'' have the same color. In this way, we obviously also obtain a proper coloring for the map in Figure 87.

PROBLEM 54. We again use the method of induction. The assertion obviously holds for a map with no more than five regions. We assume that the theorem holds for a map with no more than n regions, and we shall then prove it for a map with $n + 1$ regions.

By Problem 52, there exists a region S with fewer than six boundaries; that is, S is adjacent to no more than five other regions. If the number of neighbors of S does not exceed four, we can apply the same reasoning as in the preceding problem. It remains only to consider the case that S has exactly five neighbors. Of these five regions neighboring S there exist two, say S' and S'' (Fig. 90), that do not border on each other (Problem 51). We now erase the boundaries between S and S' and between S and S''. We obtain a new map, which consists of $n - 1$ regions and in which one large region $\overline{S}$ replaces the former regions S, S', and S''. Let us color this map properly with five colors (this is possible by assumption). We now replace the erased boundaries and get back the original map. Let S' and S'' retain the color that $\overline{S}$ had. (Since they do not adjoin one another, this is possible without spoiling the proper coloring.) The region S is then surrounded by five regions, of which two, S' and S'', have the same color; that is, S is adjacent to regions of only four different colors, and we can color it with the fifth color, which is different from the colors of its neighbors.

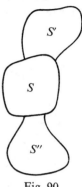

Fig. 90

PROBLEM 55. The smallest possible multiplicity of the vertices is 3. Hence, it follows from Problem 16 that

$$3v \leq 2b.$$

By Euler's theorem, $v = b + 2 - r$, from which it follows that

$$3(b + 2 - r) \leq 2b,$$
$$3b + 6 - 3r \leq 2b,$$
$$b \leq 3r - 6,$$

q.e.d.

PROBLEM 56. We wish to show that, in a connected map with less than twelve regions, there exists a region with fewer than five boundaries. We assume the contrary to be true. Then, by Problem 19,

(1) $$5r \leq 2b.$$

Since, by assumption, the multiplicity of each vertex is not less than 3, it follows from Problem 16 that

(2) $$3v \leq 2b.$$

Multiplying (1) by 3 and (2) by 5 and adding, we obtain

$$15(v + r) \leq 16b.$$

We now replace $v + r$ by $b + 2$ and get

$$15(b + 2) \leq 16b, \quad 15b + 30 \leq 16b, \quad b \geq 30.$$

On the other hand, it follows from Problem 55 that, for $r < 12$, the inequality $b < 3 \cdot 12 - 6 = 30$ is satisfied. We obtain a contradiction. Consequently, among the regions of our map there exists a region with fewer than five boundaries.

Now we obtain the solution by induction. If the number of regions does not exceed four, the coloring is possible. Assume the theorem is valid for n regions. We now prove it for the case of $n + 1$ regions ($n + 1 < 12$).

It has been proved that a map consisting of $n + 1$ regions contains a region S with fewer than five boundaries. If the number of regions which adjoin this region is equal to one, two, or three, we remove this region and apply the same reasoning as in Problem 53. If the number of neighboring regions is equal to four, however, we reason in the following way: We have five regions, namely, the region S itself and its four neighboring regions. It cannot happen that each of these is adjacent to all of the others (Problem 51). On the other hand, each of the five regions except S borders on S. Hence, there exist two regions bordering on S that do not border on each other. After we have established this fact, the rest of the proof proceeds exactly as in Problem 54. We leave it to the reader to carry out the proof for disconnected maps.

Remark. The four-color problem is solved in this exercise for the special case that the number of regions does not exceed 11. At present, it is known that all maps with no more than 38 regions can be properly colored with four colors. For maps with more than 38 regions the four-color problem remains unsolved.

Bibliography

1. Franklin, P. "Note on the Four-Color Problem," *Journal of Mathematics and Physics,* 16 (1937), 172–184.
2. Heawood, P. J. "Map-Colour Theorem," *The Quarterly Journal of Pure and Applied Mathematics,* 24 (1890), 332–339.
3. Kempe, A. B. "On the Geographical Problem of the Four Colors," *American Journal of Mathematics,* 2 (1879), 193–200.
4. Reynolds, C. N. "On the Problem of Coloring Maps in Four Colors, I," *Annals of Mathematics,* 28 (1926–27), 477–492.
5. Veblen, O. "An Application of Modular Equations in Analysis Situs," *Annals of Mathematics,* 14 (1912), 86–94.
6. Winn, C. E. "A Case of Coloration in the Four Color Problem," *American Journal of Mathematics,* 59 (1937), 515–528.
7. Winn, C. E. "On the Minimum Number of Polygons in an Irreducible Map," *American Journal of Mathematics,* 62 (1940), 406–416.

Kempe in [3] above gave an erroneous proof of the assertion that every map in the plane can be regularly colored with four colors. Heawood [2] pointed out the error in Kempe's proof. Veblen [5] transformed the four-color problem into other equivalent assertions in the fields of projective geometry and the solution of simultaneous equations. Reynolds [4] showed that every map with at most 27 regions can be colored with four colors; Franklin [1] pushed this upper bound up to 31; and Winn [7] raised it to 35. Winn [6] also proved that four colors are sufficient to color any map with at most one region with six boundaries and no regions with more than six boundaries.

SECTION TWO
Problems in the Theory of Numbers

Note

THIS section is devoted to a single topic, and it should therefore be read in order. However, Chapters 2 and 4 (marked with asterisks) deal to some extent with side issues, and may be omitted at a first reading. In fact, Chapter 2 serves as a supplement to Chapter 1, and Chapter 4 as a supplement to Chapter 3.

1. The Arithmetic of Residue Classes

1. ARITHMETIC, MODULO *m*, OR *m*-ARITHMETIC[1]

When we add two one-digit numbers we obtain either a one-digit number, for example,

$$1 + 4 = 5, \qquad 7 + 2 = 9,$$

or a two-digit number,

$$3 + 9 = 12, \qquad 5 + 8 = 13, \qquad 7 + 9 = 16, \qquad 4 + 6 = 10.$$

We now consider only the last digit in the case of two-digit sums, and we write

$$3 + 9 = 2, \qquad 5 + 8 = 3, \qquad 7 + 9 = 6, \qquad 4 + 6 = 0.$$

With this new definition of addition, the sum of two one-digit numbers is always a one-digit number.

When we multiply two one-digit numbers we likewise obtain either a one-digit number:

$$2 \cdot 3 = 6, \qquad 1 \cdot 8 = 8, \qquad 3 \cdot 3 = 9,$$

or a two-digit number:

$$6 \cdot 7 = 42, \qquad 7 \cdot 8 = 56, \qquad 9 \cdot 9 = 81.$$

Again, we shall consider only the last digit of a two-digit product, and we write

$$6 \cdot 7 = 2, \qquad 7 \cdot 8 = 6, \qquad 9 \cdot 9 = 1.$$

With this new definition of multiplication, the product of two one-digit numbers is always a one-digit number. Though the operations introduced here are different from those that we call ordinary addition and multiplication, the formulas of ordinary algebra that contain only the symbols for addition and multiplication and an

---

[1] In Chapters 1–4, *m* represents an arbitrary natural (positive whole) number greater than 1.

arbitrary number of parentheses hold also for the new operations.[1]
In particular, the following formulas hold:

$$a + (b + c) = (a + b) + c,$$
$$a + b = b + a,$$
$$a(bc) = (ab)c,$$
$$a(b + c) = ab + ac,$$

as well as the formulas

$$(a + b)^2 = a^2 + 2ab + b^2,$$
$$(a + b)^3 = a^3 + 3a^2b + 3ab^2 + b^3,$$
$$(a + b)(c + d) = ac + bc + ad + bd,$$

and others. Hence, the use of the customary symbols cannot lead
to a misunderstanding.

In this way, we have constructed a new arithmetic, which differs
from ordinary arithmetic and yet resembles it in many respects. This
new arithmetic will be helpful in the solution of many problems in
ordinary arithmetic and algebra.

We have begun to develop the *arithmetic of residue classes,
modulo* 10, or more briefly, *10-arithmetic*. It contains only the numbers 0, 1, 2, 3, 4, 5, 6, 7, 8, 9. Let us set up the addition and multiplication tables of this 10-arithmetic:

Addition table

+	0	1	2	3	4	5	6	7	8	9
0	0	1	2	3	4	5	6	7	8	9
1	1	2	3	4	5	6	7	8	9	0
2	2	3	4	5	6	7	8	9	0	1
3	3	4	5	6	7	8	9	0	1	2
4	4	5	6	7	8	9	0	1	2	3
5	5	6	7	8	9	0	1	2	3	4
6	6	7	8	9	0	1	2	3	4	5
7	7	8	9	0	1	2	3	4	5	6
8	8	9	0	1	2	3	4	5	6	7
9	9	0	1	2	3	4	5	6	7	8

Multiplication table

·	0	1	2	3	4	5	6	7	8	9
0	0	0	0	0	0	0	0	0	0	0
1	0	1	2	3	4	5	6	7	8	9
2	0	2	4	6	8	0	2	4	6	8
3	0	3	6	9	2	5	8	1	4	7
4	0	4	8	2	6	0	4	8	2	6
5	0	5	0	5	0	5	0	5	0	5
6	0	6	2	8	4	0	6	2	8	4
7	0	7	4	1	8	5	2	9	6	3
8	0	8	6	4	2	0	8	6	4	2
9	0	9	8	7	6	5	4	3	2	1

[1] These formulas may also contain powers with integral positive exponents, since such
a power is only an abbreviation for the product of several equal factors.

We make one additional remark. Suppose that we have any equality of ordinary arithmetic which contains, besides numbers, only addition signs, multiplication signs, and parentheses, and we replace every number by its last digit. We then obtain an equation that holds in 10-arithmetic. For example, we can obtain from the equalities of ordinary arithmetic,

$$(18 + 15)(123 + 1341) = 11 \cdot 8 \cdot 549,$$
$$10 + 11 + 12 + 13 + 14 + 15 = (10 + 15)3,$$
$$2^{10} + 151 = 1175,$$

the equalities of 10-arithmetic,[1]

$$(8 + 5)(3 + 1) = 1 \cdot 8 \cdot 9,$$
$$0 + 1 + 2 + 3 + 4 + 5 = (0 + 5)3,$$
$$2^{10} + 1 = 5.$$

Problem 1. Exhibit the number 0 in 10-arithmetic as the product of two factors in all possible ways.

Problem 2. Write down all possible factorizations of 1.

Problem 3. What are the last digits of the numbers

$$6^{811}, \ 2^{1000}, \ 3^{999}?$$

In 7-arithmetic there are just seven numbers: 0, 1, 2, 3, 4, 5, 6. Addition and multiplication in 7-arithmetic are determined by the following rules. To add two numbers, one must form the sum in the ordinary sense and then replace the sum by its remainder on division by 7. To multiply two numbers, one must find their product and then replace the product by the remainder on division by 7.[2] We cite a few examples:

$$3 + 5 = 1, \qquad 4 + 6 = 3, \qquad 3 + 4 = 0,$$
$$5 \cdot 3 = 1, \qquad 3 \cdot 6 = 4, \qquad 2 \cdot 6 = 5.$$

---

[1] An exponent is not replaced by its last digit. It plays a different role from the other numbers, for it indicates how often the base of the power must be multiplied by itself.

[2] An alternative terminology is as follows: Two numbers, a and b, are said to be *congruent, modulo m*, if their difference is divisible by m; this is written $a \equiv b \pmod{m}$, that is, $3 + 5 \equiv 1 \pmod 7$, etc. The set of numbers congruent to each other, modulo m, is said to form a *residue class*.

Problem 4. Compile the addition and multiplication tables of 7-arithmetic and write out all factorizations of the numbers 0 and 1.

Problem 5. Determine the remainder when the number

$$3^{100}$$

is divided by 7.

10-arithmetic and 7-arithmetic, which we have considered, are only special cases of m-arithmetic. Let m be an arbitrary positive integer greater than one. The elements of m-arithmetic are then the numbers $0, 1, 2, \ldots, m - 1$. Addition and multiplication of these m numbers is defined in the following way: To add (multiply) two numbers, one must take the remainder of the ordinary sum (product) on division by m.

Problem 6. Compile addition and multiplication tables for 2-, 3-, 4-, and 9-arithmetic.

Problem 7. Calculate

$$1 \cdot 2 \cdot 3 \cdot 4 \cdot 5 \cdot 6 \cdot 7 \cdot 8 \cdot 9 \cdot 10;$$
$$2^{10}; \ 3^{10}; \ 4^{10}; \ 5^{10}; \ 6^{10}; \ 7^{10}; \ 8^{10}; \ 9^{10}; \ 10^{10}$$

in 11-arithmetic.

Problem 8. Determine the remainder when the number

$$2^{1000}$$

is divided by 3, 5, 11, 13.

Subtraction in m-arithmetic, as in ordinary arithmetic, is defined as the operation inverse to addition. A number x is called the difference of the numbers b and a, that is, $x = b - a$, if

$$a + x = b.$$

Thus, for example, in 7-arithmetic

$$2 - 5 = 4, \quad \text{for} \quad 5 + 4 = 2;$$
$$1 - 6 = 2, \quad \text{for} \quad 6 + 2 = 1.$$

The quickest method of determining the difference of two numbers in m-arithmetic is as follows: One calculates the ordinary difference, and if it is negative, adds m to it. For example, in 7-arithmetic we have

$$1 - 5 = -4 = 3,$$
$$2 - 3 = -1 = 6.$$

Subtraction can always be carried out in m-arithmetic, and it always gives a unique solution. The use of the sign "$-$" for subtraction in m-arithmetic is justified, since every equation that holds in ordinary algebra with the "$+$", "$-$", "$\cdot$" signs and an arbitrary number of parentheses also holds in m-arithmetic. In particular, the following formulas hold:[1]

$$-(-a) = a,$$
$$-(a + b) = -a - b,$$
$$-(a - b) = -a + b,$$
$$a + (-b) = a - b,$$

as well as the formulas

$$a^2 - b^2 = (a - b)(a + b),$$
$$(a - b)^2 = a^2 - 2ab + b^2,$$
$$(a - b)^3 = a^3 - 3a^2b + 3ab^2 - b^3,$$

and others.

The m-arithmetics can be used in ordinary arithmetic as a check for addition, subtraction, and multiplication because an equality that is valid in ordinary arithmetic can be transformed into one that is valid in m-arithmetic by replacing every number in the ordinary equality by its remainder on division by m. For example, let us verify with the help of 7-arithmetic the (supposed) equality

$$74{,}218 \cdot 21{,}363 - 81{,}835 = 1{,}585{,}446{,}299. \tag{1}$$

If one replaces each number in equation (1) by its remainder on division by 7, one obtains

$$4 \cdot 6 - 5 = 3.$$

This equality does not hold in 7-arithmetic. This means that the original equality is false.

---

[1] As usual, $-a$ means the number $0 - a$. For example, in 8-arithmetic we have $-3 = 5, 7 = -1$, etc.

9-arithmetic is the one most used for verification. For any number N yields on division by 9 the same remainder as the sum of its digits. This greatly simplifies our calculations.[1]

Let us note that a verification using m-arithmetic will not always uncover an error. For example, if we check the false "equality" (1) in 9-arithmetic, we obtain the equality, correct in 9-arithmetic,

$$4 \cdot 6 - 7 = 8.$$

For greater certainty in finding errors, we could check an equality with several m-arithmetics simultaneously, say with 7-arithmetic and 9-arithmetic.

Problem 9. Prove that the sum of the cubes of all numbers in 1001-arithmetic is equal to zero.

Problem 10. (*a*) Prove that if all numbers in m-arithmetic (for odd m) are raised to the same odd power k and the results added, one obtains zero.

(*b*) Prove that for all odd numbers m and k the sum

$$1^k + 2^k + \cdots + (m - 1)^k$$

is divisible by m in ordinary arithmetic.

2. ARITHMETIC, MODULO p, OR p-ARITHMETIC[2]

We represent the numbers 0, 1, 2, 3, 4, 5, 6 in 7-arithmetic by points, and indicate by arrows what number each number becomes when multiplied by 4. We obtain, in this way, the diagram in Figure 1.[3] We call this a *multiplication diagram* for the number 4 in 7-arithmetic.

---

[1] The numbers 10, 10^2, 10^3, . . . yield the remainder 1 on division by 9. Hence the numbers

$$N = a \cdot 10^n + b \cdot 10^{n-1} + \cdots + f \text{ and } a + b + \cdots + f$$

yield the same remainder on division by 9.

[2] In Chapters 1–4, p signifies any prime number, that is, a natural number greater than 1, and with no positive divisors besides 1 and itself.

[3] Since the positions of the points that correspond to the numbers 0, 1, 2, 3, 4, 5, 6 make no difference, we have chosen them in such a way as to avoid the use of long arrows.

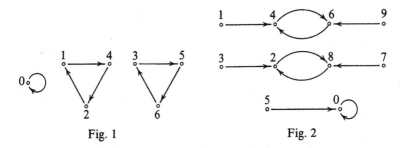

Fig. 1   Fig. 2

In exactly the same way, we construct the diagram for multiplication by 4 in 10-arithmetic (Fig. 2). From these we form the following definition.

DEFINITION. *A sequence of n numbers that appears in any diagram is called a* cycle *if an arrow can be drawn from each of these numbers to the one following, the arrow from the last number going to the first.*

The diagrams we have constructed contain the following cycles:

In Figure 1: 0; 1, 4, 2; 3, 5, 6.
In Figure 2: 0; 4, 6; 2, 8.

We point out an important difference between these two diagrams: In Figure 1, every number occurs in some cycle (and in exactly one). In Figure 2, the numbers 1, 3, 5, 7, 9, occur in no cycle.

Problem 11. Construct the diagrams for multiplication by 0, 1, 2, 3, 4, 5, 6, in 7-arithmetic, the diagrams for multiplication by 2 and 5 in 10-arithmetic, and the diagram for multiplication by 3 in 9-arithmetic.

Problem 12. Prove that if a and b are numbers in p-arithmetic[1] and $a \cdot b = 0$, either a or b is equal to zero.

Problem 13. Let a be an arbitrary number in p-arithmetic ($a \neq 0$). Prove that the diagram for multiplication by a has the following properties:
(*a*) Two arrows cannot lead to the same number.
(*b*) Some arrow leads to every number.

[1] See footnote 2 on page 76.

Problem 14. Let a and b be arbitrary numbers in p-arithmetic ($a \neq 0$). By using the preceding result, prove that in p-arithmetic exactly one number x can be found that satisfies the equation $ax = b$.

Problem 15. Let a be an arbitrary number in p-arithmetic ($a \neq 0$). Prove that
    (a) the diagram for multiplication by a consists of cycles, and
    (b) all cycles (with the exception of the zero-cycle) have the same length.

From this prove that $a^{p-1} = 1$.

Problem 16. Prove FERMAT'S THEOREM: *If p is a prime number and a is not divisible by p, then $a^{p-1} - 1$ is divisible by p.*

Division in m-arithmetic is defined as the operation inverse to multiplication. The number x is called the quotient on division of b by a (or the ratio of b to a), if $ax = b$. Division need not be possible in m-arithmetic. For example, in 10-arithmetic it is not possible to find a number x for which $4x = 5$. On the other hand, when division can be carried out, it is not always unique. For example, in 10-arithmetic $4 \cdot 8 = 2$ and $4 \cdot 3 = 2$ so that the numbers 8 and 3 are equally good candidates for the quotient of 2 on division by 4. However, it follows from Problem 14 that in p-arithmetic (p a prime number) division by an arbitrary number a ($a \neq 0$) is always possible and unique.[1] Due to this important fact, every equality of ordinary algebra that contains only the symbols "$+$", "$-$", "$\cdot$", "$\div$" is also valid in p-arithmetic. In particular, the five formulas

$$\frac{a}{b} \cdot c = \frac{ac}{b}, \qquad\qquad \frac{a}{b} \div c = \frac{a}{bc},$$

$$\frac{a}{b} \cdot \frac{c}{d} = \frac{ac}{bd}, \qquad\qquad \frac{a}{b} \div \frac{c}{d} = \frac{ad}{bc},$$

$$\frac{a}{b} + \frac{c}{d} = \frac{ad + bc}{bd}$$

remain valid, as well as others.

[1] Division of a number $a \neq 0$ by 0 is as impossible in p-arithmetic as it is in ordinary arithmetic, since there is no number x that satisfies the equation $0x = a$. It will be convenient in Chapter 3 to use the notation $\frac{a}{0} = \infty$. But one must remember that the symbol ∞ is not a number, and that one cannot operate with it as though it were.

By using *p*-arithmetic one can verify calculations with fractions in much the same way as one can verify calculations with integers in *m*-arithmetic. For example, to check the (supposed) equality

$$\frac{5}{22} + \frac{3}{17} - \frac{4}{15} = \frac{739}{5610},$$

every number in this equality is replaced by its remainder on division by the prime number[1] *p*. On division by 7, it becomes

$$\frac{5}{1} + \frac{3}{3} - \frac{4}{1} = \frac{4}{3}.$$

This equality is false in 7-arithmetic. Hence, there is an error in the original equality.

Problem 17. Compile tables of the reciprocals $1/k$ in 7-, 11-, and 13-arithmetic.

EXAMPLE. We exhibit the table of reciprocals in 5-arithmetic:

k	1	2	3	4
$\frac{1}{k}$	1	3	2	4

Show that in every *p*-arithmetic only the elements 1 and -1 are reciprocals of themselves.

Problem 18. (*a*) Prove that the product of all the nonzero elements in *p*-arithmetic is equal to -1.

(*b*) Prove WILSON'S THEOREM:[2] *If p is a prime number, then* $(p - 1)! + 1$ *is divisible by p.*

Problem 19. (The converse of Wilson's theorem). Prove that if $(m - 1)! + 1$ is divisible by *m*, then *m* is a prime number.

Problem 20. Prove that if every number in *p*-arithmetic is raised to the *k*th power and all of these powers added, one obtains either 0 or -1.

---

[1] The number *p* is to be chosen in such a way that it does not appear as a factor in any of the denominators. Otherwise, we would obtain in *p*-arithmetic an equality that contained a division by 0.

[2] Recall that *n*! is the abbreviation for the product $1 \cdot 2 \cdot 3 \cdot 4 \cdots \cdot n$ of all whole numbers from 1 to *n*. Thus, $(p - 1)! = 1 \cdot 2 \cdot 3 \cdot 4 \cdots \cdot (p - 1)$.

3. EXTRACTION OF SQUARE ROOTS IN QUADRATIC EQUATIONS

Just as before, we represent the numbers of p-arithmetic by points. We construct a diagram for squaring numbers in 5-arithmetic (Fig. 3).

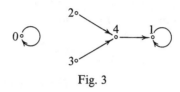

Fig. 3

Problem 21. Construct diagrams for squaring the numbers of 7-, 12-, and 24-arithmetic (the points to be arranged so that the diagrams are as simple as possible).

Problem 22. Prove that either no arrow or two arrows lead to every point (except the point 0) in the squaring diagram of p-arithmetic.[1] In other words, prove that for $a \neq 0$ the equation $x^2 = a$ in p-arithmetic has either two distinct solutions or no solution.

Problem 23. Prove that the square roots of exactly $\dfrac{p+1}{2}$ numbers can be taken in p-arithmetic (and consequently not of the remaining $\dfrac{p-1}{2}$ numbers).

If, in the diagrams that were constructed for the solution of Problem 21 (see Fig. 18 to 20), the directions of all the arrows are reversed, one obtains diagrams for the extraction of square roots in 7-, 12-, and 24-arithmetic. It is evident from these diagrams that in m-arithmetic a number a may have no square root or two or four or more different square roots. According to Problem 22, however, there are only two possibilities for square roots in p-arithmetic. Either a has no square root or a has two different square roots. (In this connection, we always assume that $a \neq 0$; 0 has only one square root, namely 0.)

---

[1] In Problems 22–25 the letter p indicates a prime number not equal to 2.

Problem 24. (*a*) Prove that in *p*-arithmetic the square root of -1 can be extracted for primes of the form[1] $p = 4k + 1$, but not for $p = 4k + 3$.

    Hint. Use Problems 15 and 18.

        (*b*) Prove that all odd prime factors of the number $a^2 + 1$ (for arbitrary *a*) have the form $4k + 1$. Every prime number of the form $4k + 1$ appears in the decomposition into prime factors of at least one number of the form $a^2 + 1$.

Problem 25. Derive the formula for the solution in *p*-arithmetic of the quadratic equation

$$ax^2 + bx + c = 0$$

(*a*, *b*, and *c* being numbers of *p*-arithmetic, $a \neq 0$). By applying this formula,[2] prove that

    if $b^2 - 4ac$ has no square root in *p*-arithmetic, then the equation has no root;

    if $b^2 - 4ac = 0$, the equation has exactly one root;

    if $b^2 - 4ac \neq 0$ and one can extract its square root, the equation has two distinct roots.

Problem 26. Solve the quadratic equations

$$5x^2 + 3x + 1 = 0,$$
$$x^2 + 3x + 4 = 0,$$
$$x^2 - 2x - 3 = 0$$

in 7-arithmetic.

Problem 27. Every quadratic equation can be brought into the form

$$x^2 + cx + d = 0$$

by dividing throughout by the coefficient of the leading term.

    The number of different quadratic equations of this form in *p*-arithmetic is p^2. Calculate how many of them have no root, one root, and two roots.

---

[1] For example, $17 = 4 \cdot 4 + 1$ is of the form $4k + 1$ and $11 = 4 \cdot 2 + 3$ is of the form $4k + 3$.

[2] The expression $b^2 - 4ac$ is called *the discriminant* of the equation $ax^2 + bx + c = 0$.

4. EXTRACTION OF CUBE ROOTS

Problem 28. Exhibit the diagram for cubing in 7-, 11-, 13-, and 17-arithmetic (see Problem 21).

If one reverses the direction of the arrows in the diagrams constructed in Problem 28 (see Figures 21 to 24 for the solution of Problem 28), one obtains diagrams for the extraction of cube roots. It is evident from the diagrams that in 11-arithmetic and 17-arithmetic there exists one and only one cube root for every number. However, in 7-arithmetic and 13-arithmetic a has either three distinct cube roots or none. ($a = 0$ is an exception. Here, the cube root has the unique value 0.)

Problem 29. Let p be a prime number of the form[1] $3k + 2$. Prove that in p-arithmetic

(a) 1 has only one cube root (namely 1) (use Problem 15);
(b) a has no more than one cube root;
(c) any a has a cube root.

Problem 30. Solve the equation

$$x^3 - 1 = 0$$

by using the equality

$$x^3 - 1 = (x - 1)(x^2 + x + 1).$$

With the help of the formulas obtained, calculate the three cube roots of 1 in 103-arithmetic.

Problem 31. Let $p > 3$. Prove that 1 in p-arithmetic has either three distinct cube roots or only one depending on whether or not -3 has a square root.

Problem 32. If the prime number p is a divisor of a number of the form

$$a^2 + 3,$$

then it either is equal to 2 or 3, or has the form $3k + 1$.

---

[1] For example, $11 = 3 \cdot 3 + 2$, $17 = 3 \cdot 5 + 2$ are numbers of such a form.

5. POLYNOMIALS AND EQUATIONS OF HIGHER DEGREE

Let us consider the polynomial

$$a_0 x^n + a_1 x^{n-1} + \cdots + a_{n-1} x + a_n \tag{1}$$

with coefficients in p-arithmetic, and $a_0 \neq 0$. In this case, n is called the *degree* of the polynomial (1). A *zero* of the polynomial

$$a_0 x^n + a_1 x^{n-1} + \cdots + a_{n-1} x + a_n$$

is a root of the equation

$$a_0 x^n + a_1 x^{n-1} + \cdots + a_{n-1} x + a_n = 0;$$

it is thus any number x_0 for which

$$a_0 x_0{}^n + a_1 x_0{}^{n-1} + \cdots + a_{n-1} x_0 + a_n = 0.$$

In the previous sections it was shown that equations (or polynomials) of the second degree have no more than two roots. We shall extend this result to polynomials of arbitrary degree.

Problem 33. Prove that if the number of zeros of a polynomial is greater than its degree, all of the coefficients are equal to zero.[1]

Problem 34. Prove that if $x^k = 1$ holds for every $x \neq 0$, k is divisible by $p - 1$.

Problem 35. Let $a \neq 0$ and $p \neq 2$. Prove that $a^{\frac{p-1}{2}} = 1$ if a has a square root in p-arithmetic, and $a^{\frac{p-1}{2}} = -1$ if it does not.

Problem 36. Prove that if a polynomial of the nth degree has n zeros, say $x_1, x_2, \ldots, x_n$, it follows that

$$a_0 x^n + a_1 x^{n-1} + \cdots + a_{n-1} x + a_n = a_0 (x - x_1)(x - x_2) \cdots (x - x_n).$$

Problem 37. Prove that in p-arithmetic

$$(x - 1)(x - 2) \cdots [x - (p - 1)] = x^{p-1} - 1.$$

From this derive Wilson's theorem (Problem 18(*b*)).

---

[1] Strictly speaking, a polynomial all of whose coefficients are 0 has no degree at all according to our definition. Here it means that the number of zeros of a polynomial cannot be greater than its degree. This assertion holds also for ordinary arithmetic.

*2. *m*-adic and *p*-adic Numbers[1]

6. THE DIVISION OF MULTIDIGIT NUMBERS USING 10-ARITHMETIC

In elementary school one learns how to carry out four arithmetic operations on multidigit numbers: addition, subtraction, multiplication, and division. The addition and multiplication of multidigit numbers can be completely reduced to the addition and multiplication of one-digit numbers. For to add or multiply two numbers, one must carry through the operations on the individual digits in a definite order, not forgetting to "carry." To subtract multidigit numbers, it suffices to be able to carry out subtraction of numbers less than twenty. The situation is entirely different in the case of division. The usual method of dividing is nothing more than systematic guessing. The simplest method of division is to estimate how large the quotient might be and multiply the conjectured quotient by the divisor. If the product is then equal to the dividend, the solution is found. If the product is larger or smaller than the dividend, then one must correspondingly decrease or increase the size of the conjectured quotient. After a finite number of guesses, we inevitably arrive at the correct solution. If the quotient has one digit, this method coincides with the usual school method. If the quotient has many digits, the rules of calculation learned in school permit the task to be broken up into individual one-digit parts. The application of 10-arithmetic allows us to divide in another way, beginning with the last digit of the dividend and the divisor, as in addition and multiplication.

We compile a table of reciprocals in 10-arithmetic:[2]

k	1	3	7	9
$\frac{1}{k}$	1	7	3	9

[1] The chapters that are marked with an asterisk contain supplementary material and may be omitted at the first reading.

[2] The numbers that lie in the same column of this table yield 1 when multiplied.

The numbers

$$2, 4, 5, 6, 8, 0$$

do not occur in this table, as they have no reciprocals. For example, when we multiply 5 by any number in 10-arithmetic, we obtain either 0 or 5, and never 1. Division cannot always be carried out in 10-arithmetic, since 10 is not a prime number. However, one can always divide by the numbers

$$1, 3, 7, \text{ and } 9.$$

To divide a number a by one of these numbers, it suffices to multiply it by the reciprocal.

Let us now consider the division of multidigit numbers. The method is best explained by considering an example. Suppose we want to find the quotient

$$74,646 \div 957.$$

Working in 10-arithmetic, we divide the last digit of the dividend by the last digit of the divisor,

$$6 \div 7 = 6 \cdot 3 = 8,$$

and we obtain the last digit of the quotient. Now, we multiply this digit by the original divisor and subtract the result from the original dividend. We obtain 66,990. We divide the last digit of this result, different from zero (i.e., 9 in this case) by 7 and obtain in 10-arithmetic,

$$9 \div 7 = 9 \cdot 3 = 7.$$

Thus, the second digit from the right of the quotient is equal to 7. We multiply the original divisor by 7 again and subtract the result from 66,990. We obtain 0, so that the process has come to an end and the quotient is 78. The scheme has the following form:

```
              78
957³)  74646
      −7656
       6699
      −6699
          0
```

The reciprocal in 10-arithmetic of the last digit of the divisor is written next to the divisor. Division by the last digit of the divisor is replaced by multiplication by the reciprocal of that number. We exhibit two more examples:

$$
\begin{array}{r}
8276 \\
74599\overline{)61730684} \\
-44754 \\
\hline
6168593 \\
-52213 \\
\hline
611638 \\
-14918 \\
\hline
59672 \\
-59672 \\
\hline
0
\end{array}
\qquad
\begin{array}{r}
739 \\
37\overline{)217} \\
-27 \\
\hline
19 \\
-9 \\
\hline
1 \\
-21 \\
-2
\end{array}
$$

In the second example there is a negative remainder. This shows that this division cannot be carried out in the whole numbers.[1] We shall return to this example in the next section.

This method of division cannot be applied directly if the divisor ends in an even number or 5. In such cases, the dividend and the divisor are divided by 2 or 5 first.[2] If necessary, the process is repeated until one finally obtains a divisor that ends in one of the digits 1, 3, 7, 9. Then division is carried out as described above.

In a number of cases, the method of division described here is simpler and more convenient than the ordinary method. However, the ordinary method has one very important advantage. If division cannot be carried out in whole numbers, the ordinary method enables one to find the quotient to any degree of accuracy. For example, it gives an accuracy of 0.1 or 0.001, which the new method does not allow.

Problem 38. Carry out the following divisions using 10-arithmetic:
(a) $37{,}233 \div 189$ (b) $36{,}408 \div 328$ (c) $851 \div 74$

7. NUMBERS WITH AN INFINITE NUMBER OF DIGITS

In elementary school one learns to subtract only in those cases in which the subtrahend is smaller than the minuend. We shall seek

---

[1] The quotient may be found as in ordinary division:
$$\frac{b}{a} = q + \frac{r}{a}; \quad \frac{217}{3} = 739 + \frac{-2000}{3} = 739 - 666\frac{2}{3} = 72\frac{1}{3}.$$

[2] To divide by 5, one can multiply by 2 and then divide by 10. Likewise, to divide by 2, one can multiply by 5 and then divide by 10. If the dividend is not divisible by 2 or 5, we obtain a decimal fraction.

to apply the same rules to the subtraction of a larger number from a smaller. For example, suppose that we are to calculate the difference $398 - 536$. We find:

$$\begin{array}{r} \dot{0}398 \\ -536 \\ \hline \overline{1}862 \end{array}$$

The calculation begins as usual with the last digits. The only novelty is that we permit ourselves to *borrow* from the zero and obtain the number -1 as the first digit of our result. (We write $\overline{1}$ instead of -1. Writing the sign "$-$" over the number makes it clear that this sign belongs only to the first digit and not to the entire number.) The result can also be represented in the following form:

$$(-1)\cdot 10^3 + 8\cdot 10^2 + 6\cdot 10 + 2 \quad \text{or} \quad -1000 + 862$$

We cite a few more examples:

$$\begin{array}{r} \overset{\cdot\;\cdot\;\cdot}{0010901} \\ -134521 \\ \hline \overline{1}876380 \end{array} \qquad \begin{array}{r} \overset{\cdot\;\cdot\;\cdot\;\cdot\;\cdot}{000001} \\ -10002 \\ \hline \overline{1}89999 \end{array} \qquad \begin{array}{r} \overset{\cdot}{01889} \\ -2354 \\ \hline \overline{1}9535 \end{array}$$

We now return to the example $217 \div 3$, considered in the previous section. We had broken off the discussion of it when we came to a negative remainder and ascertained that division in whole numbers is impossible. Now, we continue the interrupted calculation by carrying out the method of division just described.

$$\begin{array}{r} \dots 66739 \\ \overline{\quad\;} \\ 3^7)\quad 217 \\ 27 \\ \hline 19 \\ 9 \\ \hline \overset{\cdot\;\cdot}{001} \\ 21 \\ \hline \overline{0}\overline{1}8 \\ 18 \\ \hline \overline{0}\overline{1}8 \\ 18 \\ \hline \overline{1}8 \end{array}$$

If we continue the division we always obtain the remainder $(\overline{1}8)$ and the same digit (6) for the quotient. Thus, the result of the divi-

sion is a periodic whole number with infinitely many digits, namely
...6666739, or 666739 with "remainder" of $\overline{1}80000000$. We check
whether this number yields 217 when multiplied by 3,

$$
\begin{array}{cc}
6666739 & \underline{20000217} \\
\underline{\times 3} & +\overline{1}80000000 \\
20000217 & 000000217
\end{array}
$$

Thus, the new method of division, just like the ordinary method,
leads to a solution that can be written as an infinite sequence
of digits. The only difference is that this sequence is infinite to the
left rather than the right. Instead of an infinite decimal fraction we
obtain a whole number with infinitely many digits.

One can carry out addition, subtraction, and multiplication of
numbers with infinitely many digits by the same rules as those used
for numbers with finite digits. To calculate the last n digits of the
result, it is sufficient to know only the last n digits of the numbers
that are to be added, subtracted, or multiplied. We give as examples:

$$
\begin{array}{cc}
\begin{array}{r}
\ldots 100010010 \\
+\ldots 000990090 \\
\hline
\ldots 101000100
\end{array}
&
\begin{array}{r}
\ldots 175321 \\
\times \ldots 531498 \\
\hline
\ldots 402568 \\
\ldots 577889 \\
\ldots 701284 \\
\ldots 175321 \\
\ldots 525963 \\
\ldots 876605 \\
\hline
\ldots 760858
\end{array}
\end{array}
$$

$$
\begin{array}{r}
\ldots \dot{0}\dot{0}\dot{0}\dot{0}\dot{0} \\
-\ldots 00001 \\
\hline
\ldots 99999
\end{array}
$$

Let us consider division in this case. If the divisor ends in one
of the digits, 1, 3, 7, 9, then the division can be carried out by the
method described in section 6. For example,

$$
\begin{array}{r}
\ldots 40619 \\
\ldots 85143^7)\,\overline{\ldots 23517} \\
\underline{\ldots 66287} \\
\ldots 5723 \\
\underline{\ldots 5143} \\
\ldots 058 \\
\underline{\ldots 858} \\
\ldots 20 \\
\underline{\quad 2}
\end{array}
$$

The quotient is again a whole number with infinitely many digits.

It is more complicated if the divisor ends in one of the digits 0, 2, 4, 5, 6, 8. In these cases, it is not always possible to find a number, with infinitely many digits, that yields the dividend when multiplied by the divisor. For example, it is impossible to divide a number ending in 1 by 10 in the domain of integers even with infinitely many digits. A natural remedy is to consider fractional numbers with infinitely many places, for example, . . .66739.1 or . . .56429.0000017, that is, numbers that have a finite number of digits after the decimal point. Addition, subtraction, and multiplication of these numbers is carried out by the same rules as in the case of ordinary decimal fractions.

After the introduction of fractional numbers with infinitely many digits, division by 10 is easily carried out by moving the decimal point one place to the left. Likewise, division by 2 or 5 is easily carried out. It can be reduced to division by 10 followed by multiplication by 5 or 2. If we can divide a number by 2, 5, and any number a ending in 1, 3, 7, 9, it is possible to divide by any product of the form $2^m 5^n a$. For, to divide by $2^m 5^n a$ is to divide m times by 2, n times by 5, and then by a. Any number with a finite number of digits can be expressed in the form given; therefore, division by numbers with a finite number of digits is always possible. If the divisor is a number with infinitely many digits, one cannot expect it to be representable in the form $2^m 5^n a$, where a ends in one of the digits 1, 3, 7, or 9. In fact, examples (see Problem 41) show that division by numbers with infinitely many digits cannot always be carried out, even if fractional numbers with infinitely many digits are used.[1]

In conclusion, we point out a similarity between the arithmetic of numbers with infinitely many digits and the 10-arithmetic. In both of these arithmetics, addition, subtraction, and multiplication can always be carried out, and all the laws of ordinary arithmetic hold for these operations (see the formulas on pp. 2 and 5), but division cannot always be carried out.

Problem 39. Find periodic whole numbers with infinitely many digits that express the ratios 1:3, 1:7, and 1:9. Verify the results by multiplication.

---

[1] The reader may ask why one does not consider numbers with infinitely many digits after the decimal point, that is, numbers that are infinite in both directions. The fact is that for such numbers, the laws of calculation (of addition, etc.) cannot be defined in a reasonable manner. There is, so to speak, no place from which to start.

Problem 40. Prove that in the table

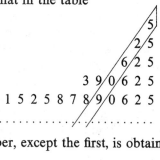

in which each number, except the first, is obtained by squaring the one above it,

(a) the (boxed) infinite number $d = \ldots 90625$ is equal to its own square, that is, $d^2 = d$;

(b) the number $e = 1 - d = \ldots 09376$ has the same property, that is, $e^2 = e$;

(c) $d \cdot e = 0$.

Hint. First prove that the last n digits of the nth and $(n + 1)$st numbers of the table coincide.

Problem 41. Prove that the number d constructed in the previous problem has no reciprocal; that is, one can find no (whole or fractional) number x with infinitely many digits such that $x \cdot d = 1$.

Problem 42. Prove that only four numbers with infinitely many digits satisfy the condition $x^2 = x$: the number 0, the number 1, and the numbers $d = \ldots 90625$ and $e = \ldots 09376$, constructed in Problem 40.

8. *m*-ADIC NUMBER SYSTEMS

If a multidigit number in the 10-adic system is written with the aid of the digits $a_n, a_{n-1}, \ldots, a_1, a_0$, it is equal to

$$a_n \cdot 10^n + a_{n-1} \cdot 10^{n-1} + \cdots + a_1 \cdot 10 + a_0.$$

The prominent role of 10 is not caused by any mathematical superiority of that number in comparison with other natural numbers. Any number $m \neq 1$ can be chosen as the basis of the numeration system. The decimal system is used simply because we are accustomed to it.

We shall consider more closely one of the m-adic systems, different from the 10-adic (decimal) system, say, the 5-adic. Here, we have only the five different digits 0, 1, 2, 3, 4. The expression 122 in the 5-adic system stands for $1 \cdot 5^2 + 2 \cdot 5 + 2$, that is, thirty-seven, and the expression 2001 stands for $2 \cdot 5^3 + 0 \cdot 5^2 + 0 \cdot 5 + 1$, that is, two hundred and fifty-one. More generally, a number that is written in the m-adic system with the digits $a_n, a_{n-1}, \ldots, a_1, a_0$ is equal to

$$a_n \cdot m^n + a_{n-1} \cdot m^{n-1} + \cdots + a_1 \cdot m + a_0$$

in the decimal system.

We now show how to change from the decimal to the 5-adic system. As an example, we consider the number 1,666 (one thousand six hundred sixty-six) and write

$$
\begin{aligned}
1666 &= 5 \cdot 333 + 1, \\
333 &= 5 \cdot 66 + 3, \\
66 &= 5 \cdot 13 + 1, \\
13 &= 5 \cdot 2 + 3, \\
2 &= 5 \cdot 0 + 2.
\end{aligned}
$$

The first of these equalities states that the number 1,666 yields the quotient 333 and remainder 1 on division by 5. In the same way, the other equalities give the result of dividing the successive quotients (333, 66, 13, 2) by 5. The number 1,666 therefore assumes the form 23,131 in the 5-adic system. Its digits are the remainders we have found. Addition, subtraction, and multiplication of multidigit numbers are carried out in the 5-adic system according to the same rules as in the decimal system. We exhibit the addition and multiplication tables for the 5-adic system below:

Addition table

+	0	1	2	3	4
0	0	1	2	3	4
1	1	2	3	4	10
2	2	3	4	10	11
3	3	4	10	11	12
4	4	10	11	12	13

Multiplication table

·	0	1	2	3	4
0	0	0	0	0	0
1	0	1	2	3	4
2	0	2	4	11	13
3	0	3	11	14	22
4	0	4	13	22	31

We cite a few examples to illustrate calculations in the 5-adic system:

```
    4302          1340              421
  + 3043        -  441           × 432
  -------        ------           -----
   12400           344            1342
                                  2313
                               + 3234
                                 ------
                                 403422
```

Everything that has been said for the 5-adic number system holds in exactly the same way for any arbitrary *m*-adic number system. One additional remark must be made concerning the digits. An *m*-adic system has *m* different digits. If $m > 10$, the usual symbols 0, 1, 2, 3, 4, 5, 6, 7, 8, 9 are not sufficient. Additional symbols must be introduced. For example, in a 12-adic system one needs two additional symbols to denote the numbers 10 and 11, say, α and β.

9. *m*-ADIC NUMBERS

We go one step further and consider, in the *m*-adic number system, not only multidigit numbers, but also numbers with infinitely many digits, that is, sequences of digits in the *m*-adic number system that extend infinitely to the left. Since the expression "numbers with infinitely many digits" does not specify *m*, one uses the more precise expression *"m-adic numbers."* Addition, subtraction, and multiplication of *m*-adic numbers differ from the operations of 10-adic numbers in the same manner as the operations with multidigit numbers in the *m*-adic number system differ from those in the 10-adic system. We illustrate examples of addition, subtraction, and multiplication of 5-adic numbers below,

```
  ...3333.1034        ...13203.1        ...43001
+ ...1111.432       - ...21403.2      × ...32104
  -----------         ----------        --------
  ...0000.0404        ...41244.4        ...32004
                                        ...001
                                        ...02
                                        ...3
                                        --------
                                        ...14104
```

We have already pointed out the similarity between the arithmetic of 10-adic numbers and 10-arithmetic. We can now say that *m*-arithmetic and the *m*-adic numbers bear the same close relation to one another.

10. p-ADIC NUMBERS

In residue class arithmetic, the p-arithmetics (p a prime number) play a special role. It is therefore to be expected that the p-adic numbers will also have particularly interesting properties. (These properties are apparent as soon as we turn to a consideration of division.) In the domain of p-adic numbers, division can always be carried out. We shall now describe how it is accomplished. If the last digit of the divisor is not equal to zero, a number can be found that is reciprocal to this last digit in the sense of residue class arithmetic. (One such number always exists.) We write this reciprocal next to the divisor and carry out the division just as we did in section 6 for division by a divisor ending in 1, 3, 7, 9. We exhibit a few examples for 7-adic numbers.

$$
\begin{array}{r}
\ldots 445 \\
\hline
3^5)\ \ldots 001 \\
-21 \\
\hline
0\overline{15} \\
-15 \\
\hline
\overline{15}
\end{array}
$$

Check
$$
\begin{array}{r}
\ldots 445 \\
\times 3 \\
\hline
\ldots 001
\end{array}
$$

$$
\begin{array}{r}
\ldots 064 \\
\hline
\ldots 346^6)\ \ldots 163 \\
-\ldots 053 \\
\hline
\ldots 11 \\
-\ldots 11 \\
\hline
\ldots
\end{array}
$$

Check
$$
\begin{array}{r}
\ldots 346 \\
\times \ldots 064 \\
\hline
\ldots 053 \\
\ldots 11 \\
+\ldots 0 \\
\hline
\ldots 163
\end{array}
$$

If the divisor ends in k zeros, it is of the form $p^k \cdot a$, where the last digit of the number a is not equal to zero. Division by such a number can be reduced to division by p^k, which is achieved by simply moving the "decimal" point k places to the left, and then dividing by a in the familiar way. For example, division in the 7-adic system of the number $\ldots 163$ by $\ldots 34,600$ is carried out in the following manner:

$$
\ldots 163 \div \ldots 34600 = (\ldots 163 \div 100) \div \ldots 346
$$
$$
= \ldots 1.63 \div \ldots 346 = \ldots 0.64.
$$

Thus, the (whole and fractional) p-adic numbers form a system in which, just as in p-arithmetic, the four operations addition, subtraction, multiplication, and division can always be carried out.

11. GEOMETRIC PROGRESSIONS

DEFINITION. *A sequence of p-adic numbers*

$$b_0, b_1, b_2, \ldots, b_n, \ldots$$

is called a geometric progression (*or* sequence) *if each of the terms, beginning with b_1, arises from the previous one upon multiplication by a constant q, the* common ratio *of the sequence.*

Thus, $b_n = b_{n-1} \cdot q$ for ($n \geq 1$). We always assume $b_0 \neq 0$ and $q \neq 0$; otherwise the progression degenerates into a sequence of zeros. In Problems 43 to 45 we consider geometric sequences whose terms are whole p-adic numbers.

Problem 43. Let a be an arbitrary p-adic number whose last digit is not equal to zero. Prove that the last digit of the number $a^{p-1} - 1$ *is* equal to zero. (If a is a number with a finite number of digits, this exercise coincides with Problem 16.)

Problem 44. Let a be a whole p-adic number whose last digit is not equal to 0. Prove that the number

$$a^{p^{k-1}(p-1)} - 1$$

ends in k zeros.

Problem 45. Prove that if we retain only the last k digits of the terms of the geometric sequence $b_0, b_1, \ldots, b_n, \ldots$ (all b_i's are p-adic numbers, and the common ratio of the sequence is not divisible by p), we obtain a periodic sequence whose length is a divisor of the number

$$p^{k-1}(p - 1).$$

In other words, prove that b_n and

$$b_{n+p^{k-1}(p-1)}$$

end in the same k digits.

12. EXTRACTION OF SQUARE ROOTS IN QUADRATIC EQUATIONS

Problem 46. Prove that the equation $x^2 = a$ either has two distinct solutions or has no solution at all in the arithmetic of p-adic numbers ($a \neq 0$).

Problem 47. Derive a formula for the solution of the quadratic equation

$$ax^2 + bx + c = 0$$

(a, b, and c are p-adic numbers, $a \neq 0$).

We now seek to find a method for extracting roots of p-adic numbers. We need to consider only the extraction of roots of whole numbers. If the number has k digits after the "decimal" point, we proceed in the following manner: We multiply it by p^{2k} (the number then becomes a whole number), then take the root, and finally divide the result by p^k.

Problem 48. Prove that the square root of a whole p-adic number is a whole p-adic number.

Problem 49. Prove that if the square root of a certain p-adic number can be taken, then the square root of its last digit in p-arithmetic can be taken in p-arithmetic.

If a whole p-adic number ends in an odd number of zeros, it is impossible to extract its square root. For, if a whole number ends in k zeros, its square will end in $2k$ zeros. If a number ends in $2k$ zeros, one can discard all of these zeros (in other words, divide by p^{2k}), then take the square root of the resulting number, and annex k zeros to the result (multiply by p^k). Hence, the extraction of square roots of arbitrary p-adic numbers can be reduced to the extraction of square roots of whole p-adic numbers whose last digit is not equal to zero.

We now show how to extract the square root of a p-adic number when the last digit of this number is different from zero and has a square root in p-arithmetic (the case $p = 2$ is excluded). First, we learn how to extract the square root with an accuracy of n digits. We call the p-adic number B_n a *square root to an accuracy of n digits* of the p-adic number A, if B_n^2 and A have their last n digits in common.

Problem 50. Verify that in triadic arithmetic the number 201 is a square root of the number ...112,101 to an accuracy of 3 digits.[1]

---

[1] We write 201 instead of ... 000201. Every number with a finite number of digits in the m-adic system can in this way be considered to be an m-adic number.

Suppose that we have already found B_n. We now show how B_{n+1}, a square root of A to an accuracy of $n + 1$ digits, may be found. We expand B_{n+1} in the form[1]

$$B_{n+1} = B_n + x \cdot 10^n,$$

where x is a one-digit p-adic number ($x \cdot 10^n = x\underbrace{00\ldots00}_{n \text{ zeros}}$).

If we square B_{n+1}, we obtain

$$B_{n+1}{}^2 = B_n{}^2 + 2B_n \cdot x \cdot 10^n + x^2 \cdot 10^{2n},$$

from which it follows that

$$2B_n \cdot x \cdot 10^n = B_{n+1}{}^2 - B_n{}^2 - x^2 \cdot 10^{2n}. \tag{1}$$

We must choose x so that the last $n + 1$ digits of $B_{n+1}{}^2$ coincide with those of A. From formula (1), it is necessary and sufficient that the numbers

$$2 \cdot B_n \cdot x \cdot 10^n$$

and

$$A - B_n{}^2 - x^2 \cdot 10^{2n}$$

have the same last $n + 1$ digits, or that this should hold for the numbers $2 \cdot B_n \cdot x \cdot 10^n$ and $A - B_n{}^2$ (since $x^2 \cdot 10^{2n}$ ends in $2n$ zeros).

Let the last digit of the number B_n be b. (Since by assumption the last digit of A is not equal to zero, it follows that $b \neq 0$.) Accordingly, the last digit of the number $2B_n$ is equal to $c = 2b$ (the product of 2 and b in p-arithmetic), where $c \neq 0$, since $p \neq 2$ and $b \neq 0$. In this case, c is the $(n + 1)$st digit from the right in the number $2B_n \cdot 10^n$. The $(n + 1)$st digit of the number $2B_n \cdot 10^n \cdot x$ is cx (the product of c and x in p-arithmetic). If we denote the $(n + 1)$st digit of the number $A - B_n{}^2$ by d, we obtain

$$cx = d;$$

and it follows that

$$x = \frac{d}{c}$$

in p-arithmetic. (Since $c \neq 0$, this division can be carried out.) We have thus found B_{n+1}.

[1] We point out that the number p is written 10 in p-adic arithmetic.

EXAMPLE. Determine the square root of the triadic number $\ldots 112{,}101$ to an accuracy of four digits. We already know the square root to an accuracy of three digits (see Problem 50). It is 201. The number we seek must have the form $(201 + x \cdot 10^3)$. We have

$$(201 + x \cdot 10^3)^2 = 201^2 + 2 \cdot 201 \cdot x \cdot 10^3 + x^2 \cdot 10^6$$
$$= 111101 + 1102000 \cdot x + x^2 \cdot 10^6.$$

The last four digits of the number $1102{,}000 \cdot x$ must be equal to the last four digits of the number

$$\ldots 112101 - 111101 - x^2 \cdot 10^6 = \ldots .001000. \qquad (2)$$

The number $1102000 \cdot x$ has (in 3-arithmetic) $2x$ as the fourth digit from the right, while the fourth digit from the right in equation (2) is 1. Hence, in 3-arithmetic,

$$2x = 1$$

and $x = 2$. We leave it to the reader to verify that $201 + 2 \cdot 10^3 = 2201$ is indeed the square root of $\ldots 112101$ to an accuracy of four digits.

We are now in a position to extract the square root of a p-adic number A. Let the last digit of A be a_1. If $\sqrt{a_1}$ can be determined in p-arithmetic, then $a_1 = b_1{}^2$. The p-adic number $B_1 = \ldots .00b_1$ is a square root of A to an accuracy of one digit. By successive applications of our process we calculate $B_2, B_3, \ldots, B_n, \ldots$.

We write these numbers one beneath another:

$$\begin{aligned}
\ldots\ldots\ldots b_1 &= B_1 \\
\ldots\ldots\, b_2 b_1 &= B_2 \\
\ldots\, b_3 b_2 b_1 &= B_3 \\
\ldots b_4 b_3 b_2 b_1 &= B_4
\end{aligned}$$

$$\ldots\ldots\ldots\ldots\ldots$$

The numbers that lie on the principal diagonal of this table form a p-adic number $B = \ldots .b_4 b_3 b_2 b_1$. It is easily seen that $B^2 = A$.

If we make use of Problem 49, we obtain the following result. *Let $p \neq 2$ and let the last digit of a p-adic number A be not equal to zero. Then one can extract the square root of A if and only if the last digit of A has a square root in p-arithmetic.*

3. Applications of m-arithmetic and p-arithmetic in Number Theory

13. THE FIBONACCI SEQUENCE

DEFINITION. *A sequence of numbers*

$$u_0, u_1, u_2, u_3, \ldots, u_n, \ldots$$

is called a Fibonacci sequence (abbreviated F-sequence), *if each of its terms beginning with u_2 is equal to the sum of the two previous terms, that is,*

$$u_n = u_{n-2} + u_{n-1} \text{ (for } n \geq 2).$$

Hence, an F-sequence is completely determined if one gives its first two terms u_0 and u_1.

By different choices of the numbers u_0 and u_1 we obtain different Fibonacci sequences. The sequence beginning with the numbers 0 and 1, that is, the sequence

$$0, 1, 1, 2, 3, 5, 8, 13, 21, 34, 55, 89, 144, 233, \ldots,$$

plays a special role among these sequences.

We agree to denote this sequence by F^0 and its terms by the letters $a_0, a_1, a_2, \ldots$:

$$a_0 = 0, a_1 = 1, a_2 = 1, a_3 = 2, a_4 = 3, a_5 = 5, a_6 = 8, \ldots.$$

Problem 51. Let $u_0, u_1, u_2, \ldots, u_n, \ldots$ be an arbitrary Fibonacci sequence. Prove that the nth term of the sequence can be expressed in terms of u_0 and u_1 by the formula $u_n = a_{n-1}u_0 + a_nu_1$. (One must assume $a_{-1} = 1$ for the formula to be valid when $n = 0$.)

Problem 52. Prove the formula

$$a_{n+m-1} = a_{n-1}a_{m-1} + a_na_m.$$

Problem 53. Prove that the sum of the squares of two neighboring numbers of the sequence F^0 (i.e., $a_{n-1}^2 + a_n^2$) again belongs to the sequence F^0.

Problem 54. Prove that if one chooses three successive terms a_{n-1}, a_n, a_{n+1} from the sequence F^0, multiplies the outer terms together, and subtracts from this product the square of the middle term, one obtains 1 or -1. Prove that for an arbitrary Fibonacci sequence (which need not begin with the numbers 0 and 1) the expression $u_{n-1} \cdot u_{n+1} - u_n^2$ has the same absolute value for every n.

Problem 55. Prove that if four successive terms a_{n-1}, a_n, a_{n+1}, a_{n+2} are chosen from the sequence F^0 and the product of the inner terms a_n, a_{n+1} is subtracted from the product of the outer terms a_{n-1}, a_{n+2}, one obtains 1 or -1. Prove that in the case of an arbitrary Fibonacci sequence the absolute value of the expression $u_{n-1} \cdot u_{n+2} - u_n \cdot u_{n+1}$ is independent of n.

14. FIBONACCI SEQUENCES IN m-ARITHMETIC

A sequence
$$v_0, v_1, v_2, \ldots, v_n, \ldots \qquad \cdot$$
of elements in m-arithmetic (see p. 3) is called a *Fibonacci sequence in m-arithmetic* (or an F_m-sequence) if for $n \geq 2$,

$$v_n = v_{n-2} + v_{n-1}.$$

(That is, addition is to be understood in the sense of m-arithmetic.) As in ordinary arithmetic, the sequence F_m is completely determined by its first two terms v_0 and v_1. For example, the sequence F_{11} that begins with the numbers 4 and 5 reads

$$4, 5, 9, 3, 1, 4, 5, \ldots.$$

The sequence that begins with the elements 0 and 1 is particularly important. We shall denote it by F_m^0 and its terms by the letters $c_0, c_1, c_2, \ldots, c_n, \ldots$. For example, the sequence F_5^0 reads:

$$0, 1, 1, 2, 3, 0, 3, 3, 1, 4, 0, 4, 4, \ldots.$$

The formulas that were derived in Problems 51–55 are also valid for sequences in m-arithmetic, if $u_0, u_1, u_2, \ldots$ and $a_0, a_1, a_2, \ldots$ are replaced by $v_0, v_1, v_2, \ldots$ and $c_0, c_1, c_2, \ldots$, respectively.

In other words, if we replace each number of the sequence F^0

$$0, 1, 1, 2, 3, 5, 8, 13, 21, 34, \ldots$$

by its remainder on division by m, we obtain the sequence F_m^0.

15. THE DISTRIBUTION OF THE NUMBERS DIVISIBLE BY m IN A FIBONACCI SEQUENCE

Problem 56. Show that the last digits of the numbers of the sequence F^0 recur periodically. What is the length of the period?

Problem 57. (Generalization of Problem 56.) Prove that all F_m-sequences are periodic, and that the length of the period is at most m^2.

Problem 58.[1] Let m be an arbitrary whole number. Prove that in the sequence F^0

$$0, 1, 1, 2, 3, 5, 8, 13, 21, 34, \ldots$$

there are infinitely many numbers divisible by m.

If n elements of m-arithmetic are arranged on a circle at equal distances from one another, and if, going clockwise, every term of the sequence is the sum of the two terms which precede it, then we call this system a *Fibonacci circular sequence* in m-arithmetic (abbreviated $\breve{F}_m$, which reads "F_m bow"). Examples of $\breve{F}_5$-sequences are shown in Figure 4.

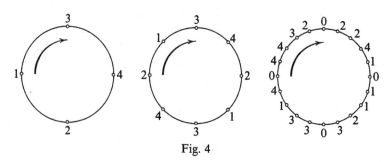

Fig. 4

We call $\breve{F}_m$ a *sequence without repetitions* if it cannot be made to coincide with itself by any rotation through an angle greater than $0°$ and less than $360°$.

---

[1] In the second round of the IX. Moscow Mathematical Olympiad for the 9th and 10th grades the following special case of this exercise was given:

Prove that among the first $10^8 + 1$ terms of the Fibonacci sequence

$$0, 1, 1, 2, 3, 5, 8, 13, 21, 34, \ldots$$

numbers occur that end in four zeros.

In Figure 4 the first and third sequences are sequences without repetitions. The second sequence is a sequence with repetitions, since it coincides with itself when rotated through 180°. If the same pair of neighboring elements occurs twice in an $\breve{F}_m$, then $\breve{F}_m$ is a sequence with repetitions. For if a sequence has the form shown in Figure 5 and $x = u$, $y = v$, it can be made to coincide with itself by a rotation carrying x into u and y into v.

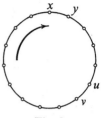

Fig. 5

According to Problem 57, every F_m-sequence is periodic. Hence, it is sufficient for the investigation of such sequences to consider only the terms forming the first period. If the numbers are written on a circle in clockwise order, one obtains an $\breve{F}_m$-sequence without repetitions.

Problem 59. Write out the circular sequences

$$\breve{F}_2{}^0, \ \breve{F}_3{}^0, \ \breve{F}_4{}^0, \ \breve{F}_5{}^0, \ \breve{F}_6{}^0, \ \breve{F}_7{}^0, \ \breve{F}_8{}^0, \ \breve{F}_9{}^0, \ \breve{F}_{10}{}^0, \ \breve{F}_{11}{}^0$$

(the $\breve{F}_m$-sequence without repetitions that contains the pair 0,1 is denoted by the symbol $\breve{F}_m{}^0$).

Problem 60. Let x and y be two elements of the sequence $\breve{F}_m$. Prove that if the sequence $\breve{F}_m$ contains a zero that has the same distance from x and y (see Fig. 6),[1] then either $x + y = 0$ or $x - y = 0$.

Problem 61. The three elements x, y, z of the sequence $\breve{F}_m$ are so arranged that the distance between x and y is equal to the distance between y and z (see Fig. 7). Prove that $x = y = 0$ implies $z = 0$.

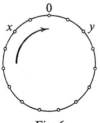

Fig. 6

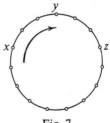

Fig. 7

[1] In the sense that the arcs $\widehat{x0}$ and $\widehat{y0}$ contain the same number of terms of the sequence.

Problem 62. Prove that the zeros contained in the sequence F_m divide the sequence into parts of equal length.

Problem 63. Prove that in the sequence F^0

$$0, 1, 1, 2, 3, 5, 8, 13, \ldots$$

the numbers divisible by m lie at equal distances from one another.

Problem 64. Prove that if a circular sequence $\breve{F}_m$ without repetitions contains zeros and consists of an odd number of elements, then the number of elements is equal to three.

Problem 65. Prove that the number of zeros contained in a circular sequence $\breve{F}_m$ without repetitions is equal to 0, 1, 2, or 4. (Thus the number of zeros in a period of an F_m-sequence is equal to 0, 1, 2, or 4.)

16. THE FIBONACCI AND GEOMETRIC SEQUENCES

DEFINITION. *A sequence of elements*

$$b_0, b_1, b_2, \ldots, b_n, \ldots$$

in m-arithmetic is called a geometric progression (sequence) *if each of its terms, beginning with b_1, is obtained by multiplying the previous term of the sequence by a constant q, the* common ratio *of the sequence.*

For example, we have in 7-arithmetic for $b_0 = 5$, $q = 3$ the geometric sequence

$$5, 1, 3, 2, 6, 4, 5, 1, 3, 2, 6, 4, 5, \ldots;$$

in 12-arithmetic for $b_0 = 1$, $q = 2$ the geometric sequence

$$1, 2, 4, 8, 4, 8, 4, \ldots.$$

The general term b_n of a geometric sequence can be expressed in terms of the initial term b_0 and the common ratio q by the formula $b_n = b_0 q^n$, which can be derived as in elementary mathematics.

We shall here consider only geometric sequences for which $b_0 \neq 0$ and $q \neq 0$ (otherwise the sequence degenerates into a sequence of zeros).

Problem 66. The periodic sequence

$$1, 4, 5, 9, 3, 1, 4, 5, 9, 3, 1, 4, 5, 9, 3, \ldots \qquad (1)$$

of elements in 11-arithmetic is at the same time a geometric sequence and a *Fibonacci sequence*. The sequence

$$b, 4b, 5b, 9b, 3b, \quad b, 4b, 5b, 9b, 3b, \quad b, 4b, 5b, 9b, 3b, \ldots ,$$

obtained from sequence (1) by multiplication by an arbitrary element b in 11-arithmetic has the same property.

(*a*) Find all sequences in 11-arithmetic that are at the same time geometric sequences and Fibonacci sequences.

(*b*) Prove that there are no such sequences in 7-arithmetic.

Problem 67. Write the sequence $F_{11}{}^0$

$$0, 1, 1, 2, 3, 5, 8, 2, 10, 1, 0, 1, 1, 2, 3, 5, 8, 2, 10, 1, \ldots$$

as the sum of two Fibonacci sequences in 11-arithmetic, each of which is also a geometric sequence.[1]

Problem 68. Prove that:

(*a*) If $\sqrt{5}$ does not exist in p-arithmetic,[2] then in this arithmetic no sequence can be constructed that is at the same time a geometric sequence and an F-sequence.

(*b*) If $\sqrt{5}$ exists in p-arithmetic, there exist in this arithmetic F-sequences that are also geometric sequences, and an arbitrary F-sequence can be represented as the sum of two such sequences.

Problem 69. If $\sqrt{5}$ exists in p-arithmetic and

$$v_0, v_1, v_2, \ldots, v_n, \ldots$$

is an arbitrary F-sequence of this arithmetic, prove that

$$v_{p-1} = v_0, \ v_p = v_1, \ v_{p+1} = v_2, \ldots, v_{k+p-1} = v_k, \ldots .$$

Problem 70. Prove that if $\sqrt{5}$ exists in p-arithmetic, then the number of elements of each circular Fibonacci sequence without repetitions is a divisor of the number $p - 1$.

---

[1] That is, find two sequences $b_0, b_1, b_2, \ldots, b_n, \ldots$ and $b_0', b_1', b_2', \ldots, b_n', \ldots$, that are at the same time geometric and Fibonacci sequences, such that

$$c_0 = b_0 + b_0', \ c_1 = b_1 + b_1', \ c_2 = b_2 + b_2', \ldots, c_n = b_n + b_n', \ldots$$

[2] In Problems 68–70, p denotes a prime number different from 2 and 5.

17. F_p-SEQUENCES

Let

$$v_0, v_1, v_2, \ldots, v_n, \ldots$$

be an arbitrary F_p-sequence. We construct the sequence

$$t_1 = \frac{v_1}{v_0}, \qquad t_2 = \frac{v_2}{v_1}, \qquad \ldots, \qquad t_n = \frac{v_n}{v_{n-1}}, \qquad \ldots,$$

and call it the *quotient sequence* of the sequence $v_0, v_1, v_2, \ldots, v_n \ldots$. If some of the numbers $v_0, v_1, v_2, \ldots, v_n, \ldots$ are equal to zero,[1] the quotient sequence contains besides numbers in p-arithmetic the symbol ∞. (If any two neighboring elements of the sequence $v_0, v_1, v_2, \ldots, v_n, \ldots$ are equal to zero, all elements of the sequence are equal to zero; we exclude this case from the discussion.)

Problem 71. Calculate the first 15 terms of the quotient sequences that correspond to the sequences

$$F_2{}^0, F_3{}^0, F_5{}^0, F_7{}^0, F_{11}{}^0, F_{13}{}^0.$$

Problem 72. (*a*) Prove that successive terms of the quotient sequence that corresponds to an arbitrary Fibonacci sequence satisfy

$$t_n = 1 + \frac{1}{t_{n-1}}.$$

By using this formula, prove that the sequence

$$t_1, t_2, \ldots, t_n, \ldots$$

is periodic and that no number appears twice in the course of a period.

(*b*) From part (*a*) derive the fact that for every prime number p the Fibonacci sequence

$$0, 1, 1, 2, 3, 5, 8, 13, 21, 34, 55, 89, 144, \ldots$$

contains infinitely many numbers that are divisible by p, and that they are distributed at equal intervals.

[1] See footnote 1, p. 78.

Problem 73. Let

$$v_0, v_1, v_2, \ldots$$

and

$$v_0', v_1', v_2', \ldots$$

be two arbitrary F_p-sequences and

$$t_1 = \frac{v_1}{v_0}, \ t_2 = \frac{v_2}{v_1}, \cdots, t_n = \frac{v_n}{v_{n-1}}, \cdots \tag{2}$$

and

$$t_1' = \frac{v_1'}{v_0'}, \ t_2' = \frac{v_2'}{v_1'}, \cdots, t_n' = \frac{v_n'}{v_{n-1}'}, \cdots \tag{3}$$

the quotient sequences corresponding to them. Prove that if (2) and (3) have an element in common (that is, if the equality $t_k = t_l'$ holds for some k and l), then they consist of the same elements; that is to say, every element of (2) occurs in (3), and conversely.

Problem 74. Let

$$t_1 = \frac{v_1}{v_0}, \ t_2 = \frac{v_2}{v_1}, \cdots, t_n = \frac{v_n}{v_{n-1}}, \cdots \tag{4}$$

and

$$t_1' = \frac{v_1'}{v_0'}, \ t_2' = \frac{v_2'}{v_1'}, \cdots, t_n' = \frac{v_n'}{v_{n-1}'}, \cdots \tag{5}$$

be quotient sequences that correspond to two arbitrary F_p-sequences. Prove that if neither t_1 nor t_1' satisfies the equation $x^2 - x - 1 = 0$, then the periods of (4) and (5) have the same length.

Hint. Consider the quotient sequence

$$\bar{t}_1 = \frac{c_1}{c_0} = \infty, \ \bar{t}_2 = \frac{c_2}{c_1} = 1, \ldots \tag{6}$$

that corresponds to the sequence

$$c_0 = 0, c_1 = 1, c_2 = 1, \ldots$$

and derive from this the formula

$$t_{n+1} = \frac{\bar{t}_n + (1 + \bar{t}_n)t_1}{1 + \bar{t}_n t_1},$$

which expresses the elements of the arbitrary quotient sequence (4) in terms of the elements of the special sequence (6).

Problem 75. (*a*) Let *r* be the length of the period of the quotient sequence of an arbitrary F_p-sequence (provided $p \neq 2$ and $p \neq 5$). Prove that

(1) If $\sqrt{5}$ does not exist in *p*-arithmetic, then *r* is a divisor of the number $p + 1$.

(2) If $\sqrt{5}$ exists in *p*-arithmetic, then *r* is a divisor of the number $p - 1$.

(*b*) Prove that for all prime numbers *p* different from 2 and 5, either a_{p+1} or a_{p-1} in the Fibonacci sequence

$$a_0 = 0, \ a_1 = 1, \ a_2 = 1, \ a_3 = 2, \ a_4 = 3, \ a_5 = 5, \ a_6 = 8, \ldots$$

is divisible by *p* in ordinary arithmetic. The first case occurs if $\sqrt{5}$ does not exist in *p*-arithmetic; the second,[1] if $\sqrt{5}$ does exist (compare with Problem 69).

18. PASCAL'S TRIANGLE

Pascal's triangle is the following infinite table of numbers:

```
                1
              1   1
            1   2   1
          1   3   3   1
        1   4   6   4   1
      1   5  10  10   5   1
    1   6  15  20  15   6   1
 . . . . . . . . . . . . . . . . . . . . . . . . . . . .
```

Every number of this table is equal to the sum of the two numbers that lie above it to the right and left. Pascal's triangle is symmetric relative to its vertical axis. We number the rows of Pascal's triangle as follows:

```
        1                      zeroth row
      1   1                    first row
    1   2   1                  second row
  . . . . . . .              . . . . . . . . . . .
```

---

[1] One can prove that $\sqrt{5}$ exists in *p*-arithmetic ($p \neq 2$ and $p \neq 5$) if and only if *p* has the form $5k \pm 1$.

The number of Pascal's triangle that lies in the nth row and at the kth place from the left, numbered from 0, is denoted here by[1]

$$C_n^k$$

(for example, $C_5^2 = 10$). By the definition of Pascal's triangle

$$C_n^k + C_n^{k+1} = C_{n+1}^{k+1}.$$

Pascal's triangle can thus be written in the form

$$
\begin{array}{ccccccc}
 & & & C_0^0 & & & \\
 & & C_1^0 & & C_1^1 & & \\
 & C_2^0 & & C_2^1 & & C_2^2 & \\
C_3^0 & & C_3^1 & & C_3^2 & & C_3^3
\end{array}
$$

. . . .

By symmetry it follows that

$$C_n^k = C_n^{n-k}.$$

We now investigate the divisibility properties of the numbers of Pascal's triangle. To this end we consider Pascal's triangle in m-arithmetic, which is defined in exactly the same way, but whose numbers are elements of m-arithmetic. If $m = 6$, Pascal's triangle has the form

$$
\begin{array}{ccccccc}
 & & & 1 & & & \\
 & & 1 & & 1 & & \\
 & & 1 & 2 & 1 & & \\
 & 1 & 3 & & 3 & 1 & \\
 1 & 4 & 0 & & 4 & 1 & \\
1 & 5 & 4 & 4 & 5 & 1 & \\
1 & 0 & 3 & 2 & 3 & 0 & 1
\end{array}
$$

. . . .

We shall denote the numbers in Pascal's triangle in m-arithmetic by P_n^k, where P_n^k is the remainder of C_n^k on division by m.

Problem 76. Prove that in the nth row of Pascal's triangle in 2-arithmetic all numbers except those on the rim (the zeroth and the nth number) are equal to zero if and only if $n = 2^k$.

---

[1] Alternative notations are $_nC_k$ or $\binom{n}{k}$.

We exhibit Pascal's triangle in 3-arithmetic:

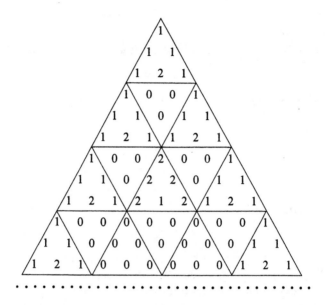

We see that (at least in those parts of the diagram that are visible) it consists of three types of triangles with point directed upward:

(We shall call these triangles *elementary* triangles.) Triangles with point directed downward contain only zeros and do not interest us here. One can introduce addition of the elementary triangles in a natural way. The sum of two triangles is the triangle whose numbers are the sums of the corresponding numbers of the two given triangles. Hence, as is obvious from the diagram, every elementary triangle is the sum of the two elementary triangles lying diagonally above it.

Pascal's triangle in 3-arithmetic is thus constructed from the elementary triangles in the same way as from the numbers. This fact, which establishes a certain periodicity in the structure of Pascal's triangle, holds in every m-arithmetic. It is more precisely formulated in Problem 77 (see below).

Let us generalize somewhat the concept of Pascal's triangle and consider triangles bordered not necessarily by 1, but by an arbitrary number a:

$$a$$
$$a \quad a$$
$$a \quad 2a \quad a$$
$$a \quad 3a \quad 3a \quad a$$

. . . .

Such a triangle is obtained from the ordinary Pascal's triangle by multiplying all of its elements by the number a. In what follows, we shall be interested not in the whole triangle, but only in the part that lies above a certain row. One can add such triangles (if they have the same number of rows) and one can multiply them by constants:

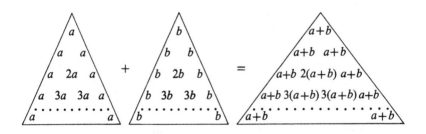

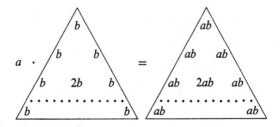

Problem 77. Let all numbers of the sth row of Pascal's triangle in m-arithmetic, with the exception of those at the edges, be equal to zero. Prove that in this case the triangle has the form represented in Figure 8. Each of the triangles $\triangle_n^k$ into which the original

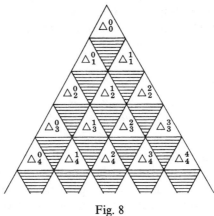

Fig. 8

triangle is decomposed has s rows. The shaded triangles contain only zeros. The triangles $\triangle_n^k$ satisfy the following equalities:

(a) $\triangle_n^{k-1} + \triangle_n^k = \triangle_{n+1}^k$,

(b) $\triangle_n^k = P_n^k \cdot \triangle_0^0$.

Problem 78. Let all the numbers of the sth row of Pascal's triangle in m-arithmetic except those at the ends be equal to zero. Prove that the rows numbered $s^2, s^3, \ldots, s^k, \ldots$ have the same property.

Problem 79. Prove that if one raises the binomial $1 + x$ to the nth power and arranges the result in increasing powers of x, one obtains the equality

$$(1 + x)^n = C_n^0 + C_n^1 x + C_n^2 x^2 + \cdots + C_n^n x^n,$$

or, in m-arithmetic,

$$(1 + x)^n = P_n^0 + P_n^1 x + P_n^2 x^2 + \cdots + P_n^n x^n.$$

Problem 80. Prove that in p-arithmetic the pth row of Pascal's triangle consists of zeros (except for the numbers at the edges). (From this it follows by Problem 78 that the rows numbered $p^2, p^3, \ldots, p^k, \ldots$ have the same property.)

19. FRACTIONAL LINEAR FUNCTIONS

The expression $ax + b$, where a and b are arbitrary numbers, defines a linear function of x. Examples are:

$$5x + 3, \qquad x + 1, \qquad \frac{3}{4}x, \qquad x.$$

The quotient of two linear expressions defines a *fractional linear function*. Examples are the expressions:

$$\frac{5x + 3}{\frac{3}{4}x + 5}, \qquad \frac{x}{x + 1}, \qquad \frac{4x + 6}{14x + 2}.$$

The general form for a fractional linear function is

$$\frac{ax + b}{cx + d}.$$

We shall denote the value of a fractional linear function by $f(x)$. If a number n is substituted for x in $f(x)$, we obtain the number $\dfrac{an + b}{cn + d}$, which we denote by $f(n)$. The linear function is a special case of the fractional linear function, since

$$ax + b = \frac{ax + b}{0 \cdot x + 1}.$$

We shall now consider linear and fractional linear functions in p-arithmetic, that is, use only numbers of p-arithmetic and add, subtract, multiply, and divide within this arithmetic. If in $ax + b$ we substitute for x a number n of p-arithmetic, we obtain the number $an + b$, which again lies in this arithmetic. Thus, a linear function makes every number of p-arithmetic correspond to a number of the same arithmetic. For example, we obtain for $f(x) = 7x + 4$ in 11-arithmetic:

$$f(0) = 4; f(1) = 0; f(2) = 7; f(3) = 3; f(4) = 10;$$
$$f(5) = 6; f(6) = 2; f(7) = 9; f(8) = 5; f(9) = 1; f(10) = 8.$$

This can be written in the form of a table

$$\begin{pmatrix} 0 & 1 & 2 & 3 & 4 & 5 & 6 & 7 & 8 & 9 & 10 \\ 4 & 0 & 7 & 3 & 10 & 6 & 2 & 9 & 5 & 1 & 8 \end{pmatrix},$$

where all possible values of x (in p-arithmetic) appear in the upper row and under each of them in the lower row is written the corresponding value for the function, $7x + 4$. Instead of a table one can also use the diagram of Figure 9. The arrows show the number into which every number in 11-arithmetic is transformed by the function. It is obvious from the diagram that the linear function defined by $7x + 4$ in 11-arithmetic leaves the number 3 fixed (3 is called a *fixed point*); the

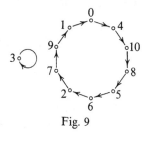

Fig. 9

other numbers form a *cycle* (for the definition of a cycle see p. 7).

In Chapter 1, section 2, we encountered a particular case of a linear function when we considered multiplication by the number a, that is, the linear function defined by ax. By repeating the arguments there set forth, we can easily ascertain that the diagram of an arbitrary linear function also falls into cycles, and that apart from cycles of length 1 (fixed points), every cycle has the same length. We shall come back to this again when we consider the same question for fractional linear functions.

We now investigate an arbitrary fractional linear function defined by

$$\frac{ax + b}{cx + d}.$$

If $ad = bc$, then

$$\frac{ax + b}{cx + d} = \frac{a}{c}\frac{c}{a}\frac{(ax + b)}{(cx + d)} = \frac{a}{c}\frac{cax + cb}{acx + ad} = \frac{a}{c},$$

so that the expression $\dfrac{ax + b}{cx + d}$ is independent of x. We shall therefore always assume $ad \neq bc$.

If we substitute numbers of p-arithmetic for the variable x to obtain a value for a fractional linear function, we can no longer be sure of obtaining a number which is once more a number in p-arithmetic. In fact, upon substituting n in $\dfrac{ax + b}{cx + d}$, the denominator $(cn + d)$ may turn out to be zero, giving us the expression $\dfrac{an + b}{0}$.

(It follows from the condition $ad \neq bc$ that in this case $an + b \neq 0$.) Division by zero is, however, not possible in p-arithmetic. We therefore enlarge p-arithmetic by introducing the symbol ∞, and agreeing to consider ∞ as a value of our function.[1] Finally, we agree to consider $\frac{a}{c}$ to be the value of the fractional linear function[2] when $x = \infty$; that is, we set $f(\infty) = \frac{a}{c}$. The fractional linear function so defined takes each of the $(p + 1)$ elements $0, 1, \ldots, p - 1$, ∞ into some one of these elements. For example, the fractional linear function with $f(x) = \frac{x + 2}{x + 1}$ in 7-arithmetic takes on the following values:

$$f(0) = 2, \ f(1) = 5, \ f(2) = 6, \ f(3) = 3,$$
$$f(4) = 4, \ f(5) = 0, \ f(6) = \infty, \ f(\infty) = 1.$$

This can be written in the form of a table

$$\begin{pmatrix} 0 & 1 & 2 & 3 & 4 & 5 & 6 & \infty \\ 2 & 5 & 6 & 3 & 4 & 0 & \infty & 1 \end{pmatrix}$$

or represented in the form of a diagram (Fig. 10).

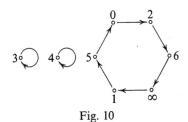

Fig. 10

From this diagram it is obvious that the fractional linear function with $f(x) = \frac{x + 2}{x + 1}$ in 7-arithmetic has the two fixed points 3 and 4; all other numbers form a single cycle.

---

[1] Compare with the footnote 1, p. 8. We point out once more that the symbol ∞ is not a number and that one cannot perform operations with it as though it were. We must therefore always consider separately the cases in which the symbol ∞ appears.

[2] It is natural to set $\frac{a}{\infty} = 0$. However,

$$f(x) = \frac{ax + b}{cx + d} = \frac{a + \dfrac{b}{x}}{c + \dfrac{d}{x}},$$

so that

$$f(\infty) = \frac{a + \dfrac{b}{\infty}}{c + \dfrac{d}{\infty}} = \frac{a + 0}{c + 0} = \frac{a}{c}.$$

Problem 81. Construct diagrams for the fractional linear functions defined by

$$\frac{4x + 1}{2x + 3}, \qquad \frac{2x + 1}{3x + 2}, \qquad \frac{3x - 1}{x + 1}$$

in 7-arithmetic.

We now consider the two fractional linear functions with

$$f(x) = \frac{ax + b}{cx + d}$$

and

$$g(x) = \frac{-dx + b}{cx - a}.$$

One sees immediately that if f transforms the number m into the number n, that is, if $f(m) = n$, then g transforms the number n back into the number m, that is, $g(n) = m$. One obtains the diagram of the function g from the diagram of the function f by simply reversing the direction of all the arrows.

This function g is said to be *inverse* to f (and then also f is inverse to g). We shall in future denote the inverse function of f by f_{-1}.

Problem 82. Let $f(x) = \dfrac{ax + b}{cx + d}$ define an arbitrary fractional linear function. Prove that one and only one arrow leads to each number in its diagram.

Problem 83. Prove that the diagram of a fractional linear function always falls into cycles and that every number of the diagram occurs in one and only one cycle (compare Problem 15).

Problem 84. If a fractional linear function transforms three numbers into themselves, then all numbers without exception are transformed into themselves.

In the investigation of a fractional linear function we are interested only in the correspondence that it sets up between the numbers of p-arithmetic, that is, in its table or diagram. In this connection the algebraic form of the function is unimportant. We can represent one and the same fractional linear function in different forms: If we multiply the numerator and denominator of such an expression by an arbitrary nonzero number, we always obtain another form defining the same function. The expressions

$$\frac{2x + 1}{7x + 5}$$

and

$$\frac{7x - 2}{8x + 1}$$

are, for example, forms for one and the same fractional linear function in 11-arithmetic (one can easily check this by writing out the corresponding tables).

Suppose we have two fractional linear functions with

$$f(x) = \frac{ax + b}{cx + d}$$

and

$$g(x) = \frac{a'x + b'}{c'x + d'}.$$

We can define a new fractional linear function by substituting $g(x)$ for x in $f(x)$:

$$f(g(x)) = \frac{a\dfrac{a'x + b'}{c'x + d'} + b}{c\dfrac{a'x + b'}{c'x + d'} + d} = \frac{(aa' + bc')x + (ab' + bd')}{(ca' + dc')x + (cb' + dd')}.$$

Problem 85. Calculate in 7-arithmetic $f(g(x))$ and $g(f(x))$ if

$$f(x) = \frac{x + 5}{5x + 1}$$

and

$$g(x) = \frac{4x + 3}{6x + 3}.$$

Problem 86. Exhibit a fractional linear function f in 7-arithmetic such that

$$f(0) = 0, \quad f(1) = 4, \quad f(4) = 2.$$

Problem 87. Let three different numbers, x_1, x_2, x_3, be given. Exhibit a fractional linear function f such that

$$f(x_1) = 0, \quad f(x_2) = 1, \quad f(x_3) = \infty.$$

Problem 88. Let three different numbers, y_1, y_2, y_3, be given. Prove that a fractional linear function f always exists such that

$$f(0) = y_1, \quad f(1) = y_2, \quad f(\infty) = y_3.$$

Problem 89. Let two triples of distinct numbers, x_1, x_2, x_3 and y_1, y_2, y_3, be given. Prove that one and only one fractional linear function f can always be found such that

$$f(x_1) = y_1, \quad f(x_2) = y_2, \quad f(x_3) = y_3.$$

Problem 90. Determine the number of distinct fractional linear functions in p-arithmetic.

We shall denote the value $f(f(x))$ by $f_2(x)$. Analogously, we introduce the fractional linear functions $f_3, f_4, \ldots, f_n$ defined by

$$f_3(x) = f(f_2(x)),$$
$$f_4(x) = f(f_3(x)),$$
$$\cdots\cdots\cdots\cdots\cdots\cdots$$
$$f_n(x) = f(f_{n-1}(x)).$$

For example,

$$f(x) = \frac{x+1}{x},$$

$$f_2(x) = \frac{\dfrac{x+1}{x} + 1}{\dfrac{x+1}{x}} = \frac{2x+1}{x+1},$$

$$f_3(x) = \frac{\dfrac{2x+1}{x+1} + 1}{\dfrac{2x+1}{x+1}} = \frac{3x+2}{2x+1}.$$

The transformations brought about by the functions f_n can be read off from the diagram of the function f. The function f causes every point of its diagram to be transformed into the point following it, that is, to move one step in the direction of the arrows. On applying the function f_2, every point in the diagram of f takes two steps, moving to the step after the next. In general, under f_n every point in the diagram of f moves n steps in its cycle (in the direction of the arrows past $n - 1$ points and to the nth place). So, one can read off diagrams for the functions f_2, f_3, ... from the diagram of the function f without calculating each time.

Problem 91. Suppose that

$$f(x) = \frac{3x - 1}{x + 1}.$$

Construct the diagrams for the functions f_2, f_3, and f_4 in 7-arithmetic.

Problem 92. Suppose that in the diagram for the function f the point x_0 lies in a cycle of length s. Prove that $f_k(x_0) = x_0$, if k is divisible by s. Conversely, if $f_k(x_0) = x_0$, then k is divisible by s.

Problem 93. Let an arbitrary fractional linear function f be given in p-arithmetic. Prove that if one of the functions $f_2, f_3, \ldots, f_k$ has even one fixed point that is not a fixed point of f, then this function leaves all points fixed.

Problem 94. Let an arbitrary fractional linear function f be given in p-arithmetic. Prove that all the cycles that compose its diagram have the same length. (We exclude the cycles that consist of only one number, that is, the fixed points of the function f; compare this with Problem 15.)

Problem 95. Let the sequence

$$x_0, \quad x_1 = f(x_0) = \frac{ax_0 + b}{cx_0 + d},$$

$$x_2 = f(x_1) = \frac{ax_1 + b}{cx_1 + d}, \quad \ldots, \quad x_k = f(x_{k-1}) = \frac{ax_{k-1} + b}{cx_{k-1} + d}, \quad \ldots$$

be given in p-arithmetic. (Thus, $x_k = f_k(x_0)$.) Prove that
  (1) If $\sqrt{(a - d)^2 + 4bc}$ does not exist in p-arithmetic, then $x_{p+1} = x_0$.
  (2) If $(a - d)^2 + 4bc = 0$, then $x_p = x_0$.
  (3) If $(a - d)^2 + 4bc \neq 0$ and $\sqrt{(a - d)^2 + 4bc}$ exists in p-arithmetic, we have $x_{p-1} = x_0$.

Problem 96. Calculate the functions f_2 and f_3 for $f(x) = \dfrac{x-3}{x+1}$ in p-arithmetic. What lengths are possible for the cycles of $f(x)$? Show that there exists a cycle of length 3.

Problem 97. (a) Let p be a prime number greater than 3. Prove that $\sqrt{-3}$ exists in p-arithmetic if and only if p has the form $3k + 1$.

(b) Prove that for any whole number a, the prime factors greater than 3 of $a^2 + 3$ all have the form $3k + 1$. Conversely, for every prime number p of the form $3k + 1$, show that an a can be found such that $a^2 + 3$ is divisible by p.

Problem 98. If

$$f(x) = \frac{x-1}{x+1},$$

calculate the functions f_2, f_3, and f_4 in p-arithmetic. What lengths are possible for its cycles? Show that there exists a cycle of length 4.

Problem 99. (a) Let p be a prime number greater than 2. Prove that $\sqrt{-1}$ exists in p-arithmetic if and only if $p = 4k + 1$.

(b) Prove that for every whole number a, all odd prime factors of the number $a^2 + 1$ have the form $4k + 1$. Every prime number of the form $4k + 1$ occurs in the decomposition into prime factors of at least one number $a^2 + 1$.

*4. Further Remarks on the Fibonacci Sequence and Pascal's Triangle[1]

20. THE APPLICATION OF p-ADIC NUMBERS TO THE FIBONACCI SEQUENCE

It was shown on page 33 that the problem of the divisibility of the numbers of a Fibonacci sequence by a prime number p is easily solved by reducing this sequence to geometric sequences, provided $\sqrt{5}$ exists in p-arithmetic. We now show that by the same considerations, numbers in the Fibonacci sequence can be found that are divisible by any arbitrary power of a prime number p^k, if we use the p-adic numbers instead of p-arithmetic.

A sequence of p-adic numbers

$$u_0, u_1, u_2, u_3, \ldots, u_n, \ldots$$

is called a Fibonacci sequence, if for $n \geq 2$

$$u_n = u_{n-2} + u_{n-1}.$$

Problem 100.[2] Determine all p-adic Fibonacci sequences that are at the same time geometric sequences. Prove that such sequences exist if and only if $\sqrt{5}$ exists in p-arithmetic.

Problem 101. Suppose that $\sqrt{5}$ exists in p-arithmetic. Prove that any p-adic F-sequence can be represented as the sum of two geometric sequences.

Problem 102. Suppose $\sqrt{5}$ exists in p-arithmetic. Prove that the last k digits of the numbers of a p-adic F-sequence are repeated periodically, and that the length of the period is a divisor of the number $p^{k-1}(p-1)$, so that the numbers labeled n and $n + p^{k-1}(p-1)$ (the numbers u_n and $u_{n+p^{k-1}(p-1)}$) have the same last k digits.

---

[1] Chapters marked with an asterisk contain supplementary material and may be omitted at the first reading.

[2] In Problems 100–103, p denotes a prime number distinct from 2 and 5.

Problem 103. Prove that if $\sqrt{5}$ exists in p-arithmetic, then in the F-sequence of ordinary whole numbers

$$0,\ 1,\ 1,\ 2,\ 3,\ 5,\ 8,\ \ldots$$

the numbers with the indices $p^{k-1}(p-1)$ (that is, $a_{p^{k-1}(p-1)}$) are divisible by p^k.

21. THE CONNECTION BETWEEN PASCAL'S TRIANGLE AND THE FIBONACCI SEQUENCE

Problem 104. Prove that the numbers of Pascal's triangle can be calculated by the formula

$$C_n^k = \frac{n!}{k!(n-k)!} \equiv \binom{n}{k}.$$

($n!$ denotes the product $1 \cdot 2 \cdot 3 \cdot \cdots \cdot (n-1) \cdot n$; it is useful to assign the value 1 to the symbol 0!, since case distinctions can then be avoided in many formulas.)

We note that for $k < p$ and $n - k < p$ the numbers $k!$ and $(n-k)!$ are distinct from zero in p-arithmetic. Hence, in this case division by $k!(n-k)!$ is possible in p-arithmetic, and the equality given in Problem 104 retains its validity in p-arithmetic as well; that is, we have[1]

$$P_n^k = \frac{n!}{k!(n-k)!}$$

Problem 105. Prove that for Pascal's triangle in p-arithmetic we have

$$P_{p-1-k}^k = (-1)^k P_{2k}^k \qquad \left(\text{for } k \leq \frac{p-1}{2}\right).$$

Problem 106. Prove that

$$P_{2n}^n = (-4)^n P_{\frac{p-1}{2}}^n \qquad \left(\text{for } n \leq \frac{p-1}{2}\right)$$

holds for Pascal's triangle in p-arithmetic.

---

[1] This equality is in p-arithmetic, so that operations on the right side are to be taken in this sense. The symbol $n!$ also is to be understood as referring to p-arithmetic, so that every factor of the product $1 \cdot 2 \cdot 3 \cdot \cdots \cdot (n-1) \cdot n$ (and also the whole product) is to be replaced by its remainder on division by p.

Numbers of the form $C_{2n}^n \left(= \binom{2n}{n} \right)$ lie on the axis of symmetry of Pascal's triangle:

$$1, \ 2, \ 6, \ 20, \ 70, \ 252, \ldots.$$

One may single out other lines, namely the diagonals, in the triangle (see Fig. 11). For example, the numbers 1, 1 lie on the third diagonal,

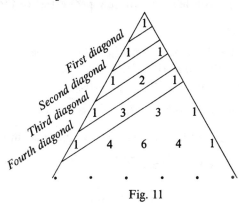

Fig. 11

the numbers 1, 2 on the fourth, and in general the numbers

$$C_{n-1}^0, \ C_{n-2}^1, \ldots, \ C_{n-1-k}^k, \ldots$$

lie on the nth diagonal (the sequence to be continued as long as the symbol C_{n-k}^{k-1} is meaningful, that is, as long as $k \le n - 1 - k$, $2k \le n - 1$, or $k \le \dfrac{n-1}{2}$).

Problem 105 shows that the kth number of the pth diagonal of a triangle in p-arithmetic is equal to the kth number of the axis of symmetry multiplied by $(-1)^{k-1}$. Problem 106 shows the connection between the axis of symmetry and the $\dfrac{p-1}{2}$th row of a triangle in p-arithmetic.

We multiply the numbers that lie on the axis of symmetry of Pascal's triangle in p-arithmetic by the numbers of an arbitrary geometric progression

$$1, \ q, \ q^2, \ \ldots$$

and add the first $\dfrac{p+1}{2}$ numbers of the sequence so constructed. We then obtain the sum

$$S = 1 + 2q + \cdots + P_{2n}^n q^n + \cdots + P_{p-1}^{\frac{p-1}{2}} q^{\frac{p-1}{2}}.$$

Problem 107. Prove that the sum S, given above, is equal to 0, $+1$, or -1 in p-arithmetic, and further, that it is equal to 0 only when $q = \frac{1}{4}$.

We now establish a connection between Pascal's triangle and the Fibonacci sequence. Aside from the fact that it is interesting in itself, this connection leads us to new properties of the Fibonacci sequence.

Problem 108. Let b_n be the sum of all the numbers that lie on the nth diagonal of Pascal's triangle. Show that $b_n = a_n$, where a_n is the nth number of the Fibonacci sequence

$$a_0 = 0, \ a_1 = 1, \ a_2 = 1, \ a_3 = 2, \ldots.$$

Problem 109. Find an expression for the pth number of an F_p-sequence $\left(\text{prove that } c_p = 5^{\frac{p-1}{2}}\right)$.

It follows from Problem 109 that $c_p = 1$, if $\sqrt{5}$ exists in p-arithmetic, and that otherwise $c_p = -1$ (Problem 35). In the first case, $c_{p-1} = 0$, as we know from Problem 75b; the first two terms of the Fibonacci sequences with the terms

$$c_0 = 0, \ c_1 = 1, \ c_2 = 1, \ldots \text{ and } c_{p-1}, \ c_p, \ c_{p+1}, \ldots$$

coincide; but then these sequences coincide completely, that is,

$$c_{p-1+k} = c_k.$$

The length of the period is in this case a divisor of the number $p - 1$. In the second case

$$c_{p+1} = 0 \text{ and } c_{p+2} = c_p + c_{p+1} = -1;$$

the first two terms differ by the factor -1; hence, the whole second sequence is obtained from the first on multiplying by -1;

$$c_{p+1+k} = -c_k,$$

from which it follows that

$$c_{2p+2} = 0 \text{ and } c_{2p+3} = -c_{p+2} = 1.$$

The length of the period is in this case a divisor of $2p + 2$.

We now explain another application of the properties of Pascal's triangle to the Fibonacci sequence.

Problem 110. Given n arbitrary numbers $d_0, \ldots d_{n-1}$, we form the sums

$$d_0^{(1)} = d_0 + d_1,$$
$$d_1^{(1)} = d_1 + d_2,$$
$$\ldots$$
$$d_{n-2}^{(1)} = d_{n-2} + d_{n-1}.$$

We thus obtain the $n - 1$ numbers

$$d_0^{(1)}, d_1^{(1)}, \ldots, d_{n-2}^{(1)}.$$

We continue this process:

$$d_0^{(2)} = d_0^{(1)} + d_1^{(1)},$$
$$\ldots$$
$$d_{n-3}^{(2)} = d_{n-3}^{(1)} + d_{n-2}^{(1)},$$

to obtain a triangular table:

d_0		d_1		$d_2 \ldots d_{n-3}$		d_{n-2}		d_{n-1}
	$d_0^{(1)}$		$d_1^{(1)}$	$\ldots$		$d_{n-3}^{(1)}$	$d_{n-2}^{(1)}$	
		$d_0^{(2)}$		$\ldots$		$d_{n-3}^{(2)}$		
			$\ldots\ldots\ldots$					
			$\ldots\ldots$					
			$d_0^{(n-1)}$					

Prove the relation

$$d_0^{(n-1)} = C_{n-1}^0 d_0 + C_{n-1}^1 d_1 + \cdots + C_{n-1}^{n-1} d_{n-1}.$$

Problem 111. Show that the sum of the squares of the numbers of a row of Pascal's triangle is again a number in Pascal's triangle.

Problem 112. Prove that in every F_p-sequence

$$v_0, v_1, v_2, \ldots v_n, \ldots$$

the equality

$$v_k + v_{k+p} = v_{k+2p}$$

holds for every k.

Problem 113. Prove that the numbers of an F_p-sequence whose indices are divisible by p again form an F_p-sequence.

22. THE DISTRIBUTION OF NUMBERS DIVISIBLE BY GIVEN NUMBERS IN A FIBONACCI SEQUENCE

We now consider the distribution of the numbers of the Fibonacci sequence that are divisible by a given number m.

Problem 114. Let a and b be whole numbers. Prove that

$$(a + b)^m = a^m + ma^{m-1}b + b^2S,$$

where S is some whole number.

Problem 115. Prove that if the kth number a_k of the F-sequence is divisible by d, then each of the differences

$$a_{kl-1} - a_{k-1}{}^l$$

and

$$a_{kl+1} - a_{k+1}{}^l$$

is divisible by d^2 for every l.

Hint. Make use of Problems 52 and 63.

Problem 116. Prove that if a_k is divisible by m^n, the difference

$$a_{k+1}{}^m - a_{k-1}{}^m$$

is divisible by m^{n+1}.

Problem 117. Prove that if a_k is divisible by m^n, a_{km} is divisible by m^{n+1}.

Problem 118. Prove that if a_k is divisible by m, all terms numbered $km^{n-1}s$ (for any s) are divisible by m^n (compare with Problem 63).

Problem 118 enables us to find the numbers that are divisible by an arbitrary power of an arbitrary given prime number p in the Fibonacci sequence. If p is different from 2 and 5, we can apply the result of Problem 75(b).

If $\sqrt{5}$ exists in p-arithmetic, a_{p-1} is divisible by p; hence, by Problem 118, all numbers of the form $a_{(p-1)p^{n-1}\cdot s}$ are divisible by p^n.

If $\sqrt{5}$ does not exist in p-arithmetic, all numbers of the form $a_{(p+1)p^{n-1}\cdot s}$ are divisible by p^n.

For $p = 2$ and $p = 5$ the indices of the numbers that are divisible by a power of p can be found directly. The Fibonacci sequence is

$$0, 1, 1, 2, 3, 5, \ldots.$$

We see that a_3 is divisible by 2 and a_5 is divisible by 5. Hence $a_{3 \cdot 2^{n-1} \cdot s}$ is divisible by 2^n, and $a_{5 \cdot 5^{n-1} \cdot s} = a_{5^n s}$ is divisible by 5^n.

We are now in a position to find numbers of the Fibonacci sequence that are divisible by m, for any number m. As an example, we shall find the indices of those numbers that are divisible by 10,000 (see footnote on p. 30). We decompose 10,000 into its factors 2^4 and 5^4. The numbers with the indices $3 \cdot 2^{4-1}s = 24s$ are divisible by 2^4, that is, the numbers whose index is divisible by 24; the numbers with the indices $5^4 t = 625t$ are divisible by 5^4, that is, all numbers whose index is divisible by 625. The numbers with an index divisible by 24 and 625 are divisible by 2^4 and 5^4, and therefore by 10,000. The indices of these numbers have the form

$$24 \cdot 625r = 15,000r.$$

We point out that our process does not in general yield *all* of the numbers of the Fibonacci sequence that are divisible by a given number m. Thus, in our example, the number with the index 12 ($a_{12} = 144$) is divisible by 2^4. Hence, all numbers with indices $12s$ are divisible by 2^4, just as all numbers with indices

$$12 \cdot 625r = 7,500r$$

are divisible by 10,000.

We consider one more example. What numbers of the Fibonacci sequence are divisible by 55,566? Factoring, we have $55,566 = 2 \cdot 3^4 \cdot 7^3$. All numbers with the index $3s$ are divisible by 2. In 3-arithmetic, $\sqrt{5}$ denotes the number $\sqrt{2}$ (2 is the remainder when 5 is divided by 3), and $\sqrt{2}$ does not exist in 3-arithmetic. Hence, the numbers with indices $(3 + 1)3^{4-1}t = 108t$ are divisible by 3^4. Likewise, in 7-arithmetic $\sqrt{5}$ does not exist, and hence the numbers with indices

$$(7 + 1)7^{3-1}q = 392q$$

are divisible by 7^3.

The numbers with indices $10,584r$ are divisible by 55,566, since 10,584 is the least common multiple of the numbers 3, 108, and 392.

The booklet by N. N. Vorobyov, *The Fibonacci Numbers,* translated into English (Boston: D. C. Heath and Company, 1963), is recommended to the reader who is particularly interested in the numbers of the Fibonacci sequence.

5. The Equation $x^2 - 5y^2 = 1$

23. SOLUTIONS OF THE EQUATION

We set ourselves the problem: To find all integral solutions of the equation[1]

$$x^2 - 5y^2 = 1, \tag{1}$$

that is, all pairs of integers a, b, for which

$$a^2 - 5b^2 = 1.$$

We put each pair of such integers a, b in correspondence with the irrational number $a + b\sqrt{5}$, which we shall call a *solution* of equation (1). Later we shall show that one obtains *all* integral solutions of equation (1) by the formula

$$x + y\sqrt{5} = \pm (9 + 4\sqrt{5})^n,$$

where n takes on all integral (not necessarily positive) values. We recommend that the reader attempt to prove this for himself before reading further.

First let us establish a few properties of the integers of the form[2] $p + q\sqrt{5}$.

Problem 119. Prove that $a + b\sqrt{5} = c + d\sqrt{5}$ implies that $a = c$ and $b = d$.

Problem 120. Verify that the product $(a + b\sqrt{5})(c + d\sqrt{5})$ can again be represented in the form $p + q\sqrt{5}$. Also, that if $a \geq 0$, $b \geq 0$, $c \geq 0$, and $d \geq 0$, then $p \geq 0$ and $q \geq 0$.

[1] Equation (1) is a special case of the equation

$$x^2 - Ay^2 = 1,$$

where A is a positive whole number with the property that $\sqrt{A}$ is irrational. The problems that we pose for equation (1) can also be solved analogously for the general equation. One can find another method of solving the equation $x^2 - Ay^2 = 1$ in the book by A. O. Gelfond, *The Solution of Equations in Integers,* translated into English (San Francisco: W. H. Freeman and Company, 1961).

[2] In Chapter 5 all letters, and in particular p and q, represent integers.

Problem 121. Prove that if

$$m + n\sqrt{5} = (a + b\sqrt{5})(c + d\sqrt{5}),$$

then

$$m - n\sqrt{5} = (a - b\sqrt{5})(c - d\sqrt{5}).$$

We shall now consider the solution of equation (1) (page 57).

Problem 122. Let $a + b\sqrt{5}$ be a solution of equation (1). Show that
  (a) $a - b\sqrt{5}$ is a solution of equation (1);
  (b) $\dfrac{1}{a + b\sqrt{5}} = a - b\sqrt{5}.$

Problem 123. Let $a + b\sqrt{5}$ and $c + d\sqrt{5}$ be solutions of equation (1). Prove that
  (a) their product

$$m + n\sqrt{5} = (a + b\sqrt{5})(c + d\sqrt{5})$$

is a solution of equation (1);
  (b) their quotient $\dfrac{a + b\sqrt{5}}{c + d\sqrt{5}}$ can be represented in the form $p + q\sqrt{5}$ and is also a solution of equation (1).

Problem 124. Verify that $9 + 4\sqrt{5}$ is a solution of equation (1) and prove that equation (1) has infinitely many different integral solutions.

Problem 125. Let $a + b\sqrt{5}$ and $c + d\sqrt{5}$ be solutions of equation (1), where $a \geq 0$, $b \geq 0$, $c \geq 0$, and $d \geq 0$. Prove that if

$$a + b\sqrt{5} < c + d\sqrt{5},$$

then

$$a < c \quad \text{and} \quad b < d.$$

Problem 126. Let $a + b\sqrt{5}$ be a solution of equation (1). Prove that
  (a) if $0 < a + b\sqrt{5}$, then $a \geq 0$;
  (b) if $1 < a + b\sqrt{5}$, then $a \geq 0$ and $b > 0$.

Problem 127. Show that there is no integral solution of equation (1) that satisfies the inequality

$$1 < a + b\sqrt{5} < 9 + 4\sqrt{5}.$$

Problem 128. Prove that all integral solutions $p + q\sqrt{5}$ of equation (1) for which $p \geq 0$ and $q \geq 0$ are given by the formula

$$p + q\sqrt{5} = (9 + 4\sqrt{5})^n,$$

where n takes on all possible whole nonnegative values.

Problem 129. Prove that one obtains all integral solutions of the equation $x^2 - 5y^2 = 1$ from the formula

$$x + y\sqrt{5} = \pm(9 + 4\sqrt{5})^n,$$

where n takes on all possible integral values

$$0, \pm 1, \pm 2, \pm 3, \pm 4, \ldots.$$

Concluding Remarks

The arithmetic of p-adic numbers as well as p-arithmetic has a great similarity to the arithmetic of ordinary rational (or real) numbers. This similarity consists in the fact that in each of these arithmetics four operations, addition, subtraction, multiplication, and division (except by zero), are defined, can always be carried out, and satisfy all of the laws of ordinary algebra. Consequently ordinary algebra can be used in p-arithmetic and in the arithmetic of p-adic numbers just as well as in ordinary arithmetic. Any arbitrary literal identity that contains only the operations of addition, subtraction, multiplication, and division, yields, on the substitution of numbers of a p-arithmetic, a valid equality in this arithmetic. Similarly, the substitution of p-adic numbers is equally valid in p-adic arithmetic, just as one obtains a valid equality in ordinary arithmetic when one substitutes real numbers.

In modern algebra, every system of elements in which the operations of addition, subtraction, multiplication, and division are defined and all the laws of ordinary algebra are satisfied is called a *field*. Accordingly, the systems of p-adic numbers as well as the p-arithmetics are *fields*. Likewise the set of all rational numbers and the set of all real numbers are *fields*.

The integers do not form a field, because division cannot always be carried out even though addition, subtraction, and multiplication do not lead out of the domain of integers. Systems of elements for which operations of addition, subtraction, and multiplication are defined and all of the laws of ordinary algebra are satisfied are called *commutative rings*. The following can be cited as examples of commutative rings: the set of all integers, the m-arithmetics for arbitrary m, the arithmetics of m-adic numbers, the set of all numbers of the form $a + b\sqrt{5}$ with integral a and b, the set of all polynomials with rational (or real) coefficients, the set of all polynomials with coefficients in p-arithmetic, and, in general, the set of all polynomials with coefficients in an arbitrary field. Obviously, every *field* is a *commutative ring*.

It turns out that all of the laws of algebra can be deduced from a small number of *fundamental laws*. These fundamental laws usually include the following:

(1) $a + b = b + a$ (commutativity of addition).

(2) $(a + b) + c = a + (b + c)$ (associativity of addition).

(3) For every a and b there exists a *difference*; that is, there exists an x such that $a + x = b$ (in other words, subtraction can always be carried out).

(4) $ab = ba$ (commutativity of multiplication).

(5) $(ab)c = a(bc)$ (associativity of multiplication).

(6) $a(b + c) = ab + ac$ (distributivity of multiplication with respect to addition).

(7) for every b and every $a \neq 0$[1] there exists a quotient; that is, there is an x such that $ax = b$ (division is possible).

Accordingly, a field can be defined as a system of elements in which the operations of addition and multiplication are defined and the properties (1) to (7) are satisfied. Analogously, one can define a commutative ring as a system of elements satisfying the properties (1) to (6).[2]

In this booklet we have encountered another class of algebraic systems, called *groups*. In a group there exists an operation which

---

[1] Strictly, one should first define 0. Pick any a in the field, and let x be such that

$$a + x = a \quad \text{(by (3))}.$$

Problem 130. Prove that x is uniquely defined; that is, if $a + x = a$ and $a + y = a$, then $x = y$.

Problem 131. Prove that for all b in a field,

$$b + x = b.$$

We write 0 instead of x.

Problem 132. Prove that for all c in the field

$$c\,0 = 0.$$

This is the reason why we cannot allow 0 to have an inverse.

[2] Systems that satisfy the fundamental laws (1) to (3) and (5) to (6) (but need not satisfy fundamental law (4)) and for which $(b + c)a = ba + ca$ holds in addition, are called simply *rings*.

we call composition. We shall denote the composite of the elements a and b by $a * b$. This operation satisfies the following laws:

I. For every triple of elements

$$(a * b) * c = a * (b * c).$$

II. There exists a *unit element*, that is, an element e such that $e * a = a$ for every a.

III. For every element a, there exists an *inverse*, that is, an element a^{-1} with the property

$$a^{-1} * a = e.$$

The product of solutions of the equation $x^2 - 5y^2 = 1$ is again a solution of this equation. If we take this product as the composite of two solutions, then they form a group (since the other properties are also satisfied). The set of fractional linear functions with coefficients in an arbitrary field also form a group, if composition is defined as the substitution:[1]

$$f(x) * g(x) = f(g(x)).$$

Finally, all of the elements different from zero in an arbitrary field form a group (the composition being taken as multiplication). Likewise do the elements of an arbitrary ring (the composition being addition). We leave it to the reader to verify that all these sets are in fact groups. (One must check the validity of the laws I, II, III.)

The concepts of group, ring, and field are fundamental concepts of modern algebra and play a prominent role in all of mathematics. A good introduction to group theory is given by W. Ledermann in *Introduction to the Theory of Finite Groups* (New York: Interscience Publishers, Inc., 1957); also by Nathan Jacobson in Chapter 1 of *Lectures in Abstract Algebra*, Vol. I (New York: Van Nostrand, 1951). For an introduction to rings and fields, the above book by Jacobson is very good.

The problems that we have placed in this booklet lie on the borderline between Algebra and Number Theory and fall within the subject known as Algebraic Number Theory. The works of the

---

[1] In this example the composite of two elements depends on their order, since in general $f(g(x))$ is not equal to $g(f(x))$ (see Problem 85).

following authors have been fundamental in the development of this theory: Carl Friedrich Gauss (1777–1855), Évariste Galois (1811–1832), Ernst Kummer (1810–1893), Leopold Kronecker (1823–1891), Egor I. Zolotarev (1847–1878), Richard Dedekind (1831–1916), and David Hilbert (1862–1943).

The theory of p-adic numbers that was dealt with in Chapter 2 was developed by Ernst Kummer and Kurt Hensel. Of the works on algebraic number theory that have appeared in the course of the last decade, we wish to mention those of N. G. Chebotarev, H. Hasse, E. Hecke, E. Landau, C. L. Siegel, and I. R. Shafarevich.

We mention the following work from the popular literature which includes some sections on algebraic number theory: H. Rademacher and O. Toeplitz, *The Enjoyment of Mathematics* (Princeton, N.J.: Princeton University Press, 1957).

Solutions to Problems

PROBLEM 1. We have

$$0 = 0 \cdot 0 = 0 \cdot 1 = 0 \cdot 2 = 0 \cdot 3 = 0 \cdot 4 = 0 \cdot 5 = 0 \cdot 6$$
$$= 0 \cdot 7 = 0 \cdot 8 = 0 \cdot 9 = 2 \cdot 5 = 4 \cdot 5 = 6 \cdot 5 = 8 \cdot 5.$$

The product of an arbitrary number with the number 0 is, just as in ordinary arithmetic, always equal to zero. However, in contrast to ordinary arithmetic, the product of two nonzero factors can also be equal to zero.

DEFINITION. Nonzero numbers, such as 2, 4, 6, 8, 5, which give zero when multiplied as indicated above, are called *divisors of zero*.

PROBLEM 2. We have

$$1 = 1 \cdot 1 = 3 \cdot 7 = 9 \cdot 9.$$

We point out that no divisor of zero occurs in these equalities, but that every other number without exception does occur.

PROBLEM 3. For a solution of the exercise it suffices to calculate the expressions 6^{811}, 2^{1000}, 3^{999} in 10-arithmetic. Here we have the equality $6^2 = 6$. If this equality is successively multiplied by 6, 6^2, 6^3, ..., we obtain

$$6^3 = 6^2, \; 6^4 = 6^3, \; 6^5 = 6^4, \ldots.$$

From this, it follows that $6^n = 6$ for arbitrary n. In particular $6^{811} = 6$. To calculate 2^{1000} we remark that $2^4 = 6$. Hence, $2^{1000} = 2^{4 \cdot 250} = 6^{250} = 6$. Finally, from the equality $3^4 = 1$ we obtain

$$3^{999} = 3^{4 \cdot 249 + 3} = (3^4)^{249} \cdot 3^3 = 3^3 = 7.$$

PROBLEM 4. Addition table

+	0	1	2	3	4	5	6
0	0	1	2	3	4	5	6
1	1	2	3	4	5	6	0
2	2	3	4	5	6	0	1
3	3	4	5	6	0	1	2
4	4	5	6	0	1	2	3
5	5	6	0	1	2	3	4
6	6	0	1	2	3	4	5

Multiplication table

·	0	1	2	3	4	5	6
0	0	0	0	0	0	0	0
1	0	1	2	3	4	5	6
2	0	2	4	6	1	3	5
3	0	3	6	2	5	1	4
4	0	4	1	5	2	6	3
5	0	5	3	1	6	4	2
6	0	6	5	4	3	2	1

The decompositions of zero and one into factors have the form

$$0 = 0 \cdot 0 = 1 \cdot 0 = 2 \cdot 0 = 3 \cdot 0 = 4 \cdot 0 = 5 \cdot 0 = 6 \cdot 0,$$
$$1 = 1 \cdot 1 = 2 \cdot 4 = 3 \cdot 5 = 6 \cdot 6.$$

We notice that in 7-arithmetic the product of two numbers is equal to zero if and only if one of the two factors is equal to zero. In this respect 7-arithmetic is like ordinary arithmetic and different from 10-arithmetic.

We now compare the multiplication table of 7-arithmetic with the multiplication table of 10-arithmetic (see p. 2). The important difference between these two tables is that in every row of the table for 7-arithmetic (except that corresponding to zero) each number occurs once and only once. In contrast, some numbers in the multiplication table for 10-arithmetic are missing in certain rows, while others occur more than once. The rows with repetitions (except the row zero) are 2, 4, 5, 6, 8. These numbers correspond to the divisors of zero (see solution of Problem 1). There are no rows with repetitions in the multiplication table for 7-arithmetic because there are no divisors of zero in this arithmetic.[1]

PROBLEM 5. We calculate 3^{100} in 7-arithmetic:

$$3^2 = 2, \ 3^3 = 6, \ 3^4 = 4, \ 3^5 = 5, \ 3^6 = 1.$$

From this it follows that

$$3^{100} = 3^{6 \cdot 16 + 4} = (3^6)^{16} \cdot 3^4 = 3^4 = 4.$$

PROBLEM 6. We exhibit only the tables for 2-arithmetic and 4-arithmetic:

2-arithmetic		4-arithmetic	

Addition table	Multiplication table	Addition table	Multiplication table

+	0	1
0	0	1
1	1	0

·	0	1
0	0	0
1	0	1

+	0	1	2	3
0	0	1	2	3
1	1	2	3	0
2	2	3	0	1
3	3	0	1	2

·	0	1	2	3
0	0	0	0	0
1	0	1	2	3
2	0	2	0	2
3	0	3	2	1

The multiplication tables of 2-arithmetic and 3-arithmetic are of the same type as for 7-arithmetic: with the exception of the row 0, there is no row with repetitions. On the other hand, the multiplication tables for 4- and 9-arithmetic, like the multiplication table for 10-arithmetic, contain rows with repetitions besides the zero row.

---

[1] Zero itself is not a *divisor of zero*.

PROBLEM 7. $1 \cdot 2 \cdot 3 \cdot 4 \cdot 5 \cdot 6 \cdot 7 \cdot 8 \cdot 9 \cdot 10 = 10$;

$$2^{10} = 3^{10} = \ldots = 10^{10} = 1.$$

PROBLEM 8. 1; 1; 1; 3.

PROBLEM 9. We must show that in 1001-arithmetic

$$1^3 + 2^3 + 3^3 + \cdots + 500^3 + 501^3 + \cdots + 998^3 + 999^3 + 1000^3 = 0. \quad (1)$$

But in 1001-arithmetic,

$$1000 = -1, \qquad\qquad 1000^3 = (-1)^3 = -1^3,$$
$$999 = -2, \qquad\qquad 999^3 = (-2)^3 = -2^3,$$
$$998 = -3, \qquad\qquad 998^3 = (-3)^3 = -3^3,$$
$$\cdots\cdots\cdots\cdots \qquad\qquad \cdots\cdots\cdots\cdots\cdots\cdots\cdots\cdots$$
$$501 = -500 \qquad\qquad 501^3 = (-500)^3 = -500^3.$$

Hence, the corresponding numbers of the sum (1) cancel each other.

PROBLEM 10. (a) See the solution of Problem 9.

(b) The expression $1^k + 2^k + \cdots + (m-1)^k$ is, according to part (a), equal to zero in m-arithmetic. Hence it is divisible by m in ordinary arithmetic.

PROBLEM 11. As examples, we exhibit the diagram for multiplication by 3 in 7-arithmetic (Fig. 12), the diagram for multiplication by 5 in 10-arithmetic (Fig. 13), and the diagram for multiplication by 3 in 9-arithmetic (Fig. 14).

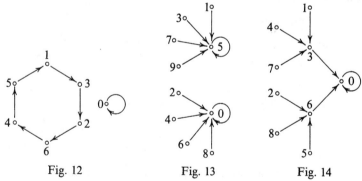

Fig. 12　　　　　　　Fig. 13　　　　　　　Fig. 14

The first diagram breaks up into cycles, whereas the other two do not.

PROBLEM 12. To find the product of the numbers a and b in p-arithmetic is to find their product in the ordinary sense and then take the remainder on division by p. Hence, a product is equal to zero in p-arithmetic if and only if the ordinary product is divisible by p. However, it is known from ordinary arithmetic that a product is divisible by a prime number p only when one of the factors is divisible by p. Hence, either a or b is divisible by p. But a and b are among the numbers $0, 1, 2, \ldots, p - 1$; therefore, either $a = 0$ or $b = 0$.

Remark. The following fact is easily proved. If, in an m-arithmetic, $ab = 0$ implies that either $a = 0$ or $b = 0$, then m is a prime number. It follows from this and Problem 12 that we can define a p-arithmetic as an m-arithmetic that contains no divisor of zero.

PROBLEM 13. (*a*) If two arrows were to lead to the number x, say from the numbers y and z, this would mean that $ay = x$ and $az = x$. If we subtract the second equation from the first, we obtain the equation $a(y - z) = 0$, which is impossible according to Problem 12, as it was assumed that $a \neq 0$ and $y \neq z$.

(*b*) Exactly one arrow leads out from each number. Consequently, the number of arrows in the diagram is equal to the number of points in the diagram, namely, p. Every arrow leads to some number. If no arrow at all were to lead to some one number, then two arrows would have to lead to the same number, which is impossible according to Problem 13(*a*).

To clarify the reasoning just carried through, we cite an example. Suppose 30 workbooks have been given out to a class of 30 students. If it is known that none of the students has been given more than one workbook, then we can conclude that *every* student has received exactly *one* workbook.

PROBLEM 14. Let us consider the diagram for multiplication by a. By Problem 13(*b*) one arrow leads to the number b from a number x. But this means exactly that $ax = b$.

PROBLEM 15. (*a*) Let b be an arbitrary number in p-arithmetic. We show that b is contained in some cycle. We start out from b on our diagram, going from number to number in the direction of the arrows. After several steps (the number of the steps can vary from 1 to p) we arrive at a point where we have already been. This point can be only the point b; otherwise our path would have the form shown in Figure 15; that is, two arrows would lead to the same number, which is impossible by Problem 13(*a*).

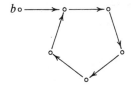

Fig. 15

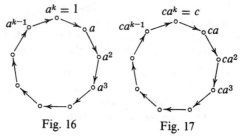

Fig. 16   Fig. 17

(*b*) Let us consider the cycle that contains the number 1, and suppose it has the form shown in Figure 16. We multiply this cycle by an arbitrary number $c \neq 0$ to obtain a new cycle of this diagram (Fig. 17), which has the same length as the original cycle. It is clear that every cycle C of the diagram, with the exception of the zero-cycle, can be obtained in this way. For if C contains the number c it suffices to multiply the cycle containing 1 by c. Hence, all cycles different from zero have the same length.

We denote the length of a cycle different from zero by k and the number of such cycles by s. Then we have $ks = p - 1$. On the other hand, $a^k = 1$ (Fig. 16). If we raise this equality to the sth power, i.e. $(a^k)^s = 1^s = 1$, we obtain

$$a^{ks} = a^{p-1} = 1.$$

PROBLEM 16. This follows from the fact that in p-arithmetic $a^{p-1} - 1 = 0$ (see Problem 15).

PROBLEM 17. The tables of reciprocals are:

in 7-arithmetic

k	1	2	3	4	5	6
$\frac{1}{k}$	1	4	5	2	3	6

in 11-arithmetic

k	1	2	3	4	5	6	7	8	9	10
$\frac{1}{k}$	1	6	4	3	9	2	8	7	5	10

in 13-arithmetic

k	1	2	3	4	5	6	7	8	9	10	11	12
$\frac{1}{k}$	1	7	9	10	8	11	2	5	3	4	6	12

Suppose x is equal to its own reciprocal, that is, $x = \frac{1}{x}$. Then $x^2 = 1$, from which it follows that $(x - 1)(x + 1) = 0$. By Problem 12, it follows that
$$x - 1 = 0 \quad \text{or} \quad x + 1 = 0, \quad \text{i.e.,}$$
$$x = 1 \quad \text{or} \quad x = -1.$$

PROBLEM 18. (a) Every number in the product $1 \cdot 2 \cdot 3 \cdot 4 \cdot 5 \cdots (p - 1)$ except 1 and $p - 1$ cancels its reciprocal (see Problem 17). Hence,

$$1 \cdot 2 \cdot 3 \cdot 4 \cdot 5 \cdots (p - 1) = 1 \cdot (p - 1) = -1.$$

(b) It follows from (a).

PROBLEM 19. We assume the contrary. Let m be a composite (not a prime) number. Then m is divisible by some number d, where $1 < d < m - 1$. Then $(m - 1)!$ is divisible by d, and $(m - 1)! + 1$, which is divisible by m, is likewise divisible by d. But then the difference

$$[(m - 1)! + 1] - (m - 1)! = 1$$

must also be divisible by d, which contradicts the assumption that $d > 1$.

PROBLEM 20. It is to be proved that the sum

$$S = 0^k + 1^k + 2^k + \cdots + (p - 1)^k \tag{1}$$

is equal to 0 or -1.

We first consider the case where the relation $a^k = 1$ holds for every number a $(a \neq 0)$ in p-arithmetic.[1] Then

$$S = 0^k + 1^k + 2^k + \cdots + (p - 1)^k = 0 + \underbrace{1 + 1 + \cdots + 1}_{p - 1 \text{ times}} = p - 1 = -1.$$

Now suppose that $a^k = 1$ does not hold for all numbers a different from zero; that is, there exists a number $b \neq 0$ such that $b^k \neq 1$. We multiply (1) by b^k and obtain

$$Sb^k = [0^k + 1^k + 2^k + \cdots + (p - 1)^k]b^k$$
$$= (0 \cdot b)^k + (1 \cdot b)^k + (2 \cdot b)^k + \cdots + [(p - 1) \cdot b]^k.$$

All of the numbers of p-arithmetic occur among the numbers $0 \cdot b$, $1 \cdot b$, $2 \cdot b, \ldots, (p - 1) \cdot b$ (for every a we have $\dfrac{a}{b} \cdot b = a$). Furthermore, every number a occurs only once, since $a = xb$ implies that $x = \dfrac{a}{b}$. Hence,

$$(0 \cdot b)^k + (1 \cdot b)^k + (2 \cdot b)^k + \cdots + [(p - 1) \cdot b]^k$$

is the sum (1), written in a different order.

$$Sb^k = S \quad \text{and} \quad S(b^k - 1) = 0.$$

Since $b^k - 1 \neq 0$, one can divide both sides of the equality by this factor and obtain $S = 0$.

---

[1] One easily sees that this is the case when k is divisible by $p - 1$. Then one can prove that there is no other possibility; that is, if $a^k = 1$ holds for every a, then k is divisible by $p - 1$ (see Problem 34).

PROBLEM 21. The diagrams for squaring are shown below; for 7-arithmetic in Figure 18, for 12-arithmetic in Figure 19, and for 24-arithmetic in Figure 20.

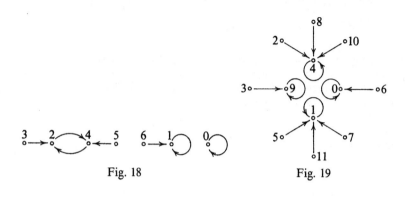

Fig. 18

Fig. 19

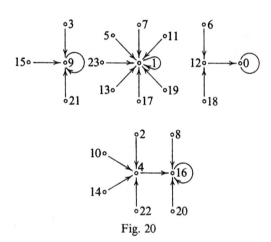

Fig. 20

PROBLEM 22. We assume that the equation

$$x^2 = a \tag{1}$$

is satisfied by some number b in p-arithmetic, that is, that

$$b^2 = a. \tag{2}$$

If we subtract equality (2) from equation (1), we obtain

$$x^2 - b^2 = 0,$$

so that

$$(x - b)(x + b) = 0.$$

If the product of two numbers in p-arithmetic is equal to zero, one of the two factors is equal to zero (Problem 12). Hence $x - b = 0$ or $x + b = 0$; that is, either $x = b$ or $x = -b$. We prove that for $a \neq 0$ these two solutions are distinct. In fact, if $b = -b$, then $b + b = 0$. If we are not dealing with 2-arithmetic, it follows that $b = 0$, and hence also $a = 0$.

PROBLEM 23. Let us consider the numbers whose square roots can be extracted in the squaring diagram in p-arithmetic, that is, points to which at least one arrow leads. Let k be the number of such points. Then, by virtue of Problem 22, one arrow leads to one of them, namely the point 0, and two arrows lead to each of the other $(k - 1)$ points. The number of arrows in the diagram is thus

$$2(k - 1) + 1.$$

On the other hand, exactly one arrow goes out from each point. But there are exactly p points, hence, p arrows in the diagram. Therefore,

$$2(k - 1) + 1 = p, \quad \text{or} \quad k = \frac{p + 1}{2}.$$

PROBLEM 24. (*a*) Let $p = 4k + 1$. By Problem 18,

$$1 \cdot 2 \cdot \cdots \cdot (p - 1) = -1.$$

We write this equality in the form

$$-1 = 1 \cdot 2 \cdot \cdots \cdot (p - 1)$$

$$= \left[1 \cdot 2 \cdot \cdots \cdot \frac{p - 3}{2} \cdot \frac{p - 1}{2} \right]$$

$$\times \left[\left(\frac{p - 1}{2} + 1 \right)\left(\frac{p - 1}{2} + 2 \right) \cdots (p - 2)(p - 1) \right]$$

$$= \left[1 \cdot 2 \cdot \cdots \cdot \frac{p - 3}{2} \cdot \frac{p - 1}{2} \right]$$

$$\times \left[\left(p - \frac{p - 1}{2} \right)\left(p - \frac{p - 3}{2} \right) \cdots (p - 2)(p - 1) \right]$$

$$= \left[1 \cdot 2 \cdot \cdots \cdot \frac{p - 3}{2} \cdot \frac{p - 1}{2} \right]\left[(-1)(-2) \cdots \left(-\frac{p - 3}{2} \right)\left(-\frac{p - 1}{2} \right) \right]$$

$$= \left[1 \cdot 2 \cdot \cdots \cdot \frac{p - 3}{2} \cdot \frac{p - 1}{2} \right]^2 (-1)^{\frac{p - 1}{2}}$$

$$= \left[1 \cdot 2 \cdot \cdots \cdot \frac{p - 3}{2} \cdot \frac{p - 1}{2} \right]^2 (-1)^{2k}$$

$$= \left[1 \cdot 2 \cdot \cdots \cdot \frac{p - 3}{2} \cdot \frac{p - 1}{2} \right]^2$$

so that the square root of -1 can be extracted when $p = 4k + 1$.

Conversely, suppose that $\sqrt{-1}$ exists, so that

$$x^2 = -1.$$

We raise this equality to the $\dfrac{p-1}{2}$th power.

$$x^{2\left(\frac{p-1}{2}\right)} = -1^{\frac{p-1}{2}}.$$

But by Problem 15,

$$x^{2\left(\frac{p-1}{2}\right)} = x^{p-1} = 1.$$

From this it follows that

$$(-1)^{\frac{p-1}{2}} = 1$$

and

$$\frac{p-1}{2} = 2k;$$

that is,

$$p = 4k + 1.$$

Thus, the square root of -1 cannot be taken when $p = 4k + 3$.

(b) Suppose that $a^2 + 1$ is divisible by p. In p-arithmetic this is to say $b^2 + 1 = 0$ (where b is the remainder on dividing a by p); that is, $\sqrt{-1}$ exists in p-arithmetic. Hence, by the previous exercise, $p = 4k + 1$. On the other hand, if $p = 4k + 1$, $\sqrt{-1}$ exists in p-arithmetic; that is, there exists an a such that $a^2 = -1$ and $a^2 + 1 = 0$. In terms of ordinary arithmetic this means that $a^2 + 1$ is divisible by p.

PROBLEM 25. The formula is derived exactly as in ordinary algebra. We use the identity

$$ax^2 + bx + c = a\left(x + \frac{b}{2a}\right)^2 + \frac{4ac - b^2}{4a}.$$

According to this identity the equation

$$ax^2 + bx + c = 0 \tag{1}$$

is equivalent to the equality

$$a\left(x + \frac{b}{2a}\right)^2 + \frac{4ac - b^2}{4a} = 0$$

or

$$\left(x + \frac{b}{2a}\right)^2 = \frac{b^2 - 4ac}{4a^2}. \tag{2}$$

It is obvious from the last equality that if equation (1) has a solution, then the square root of $b^2 - 4ac$ can be extracted. In this case one can write equality (2) in the form

$$x + \frac{b}{2a} = \frac{\pm \sqrt{b^2 - 4ac}}{2a},$$

from which it follows that

$$x = \frac{-b \pm \sqrt{b^2 - 4ac}}{2a}. \tag{3}$$

Thus, equation (1) cannot be solved if the square root of $b^2 - 4ac$ cannot be taken, and it has two solutions, which can be calculated by formula (3), if $\sqrt{b^2 - 4ac}$ exists. These two solutions are distinct in the case that $b^2 - 4ac \neq 0$, and they coincide for $b^2 - 4ac = 0$. Formula (3) does not make sense in 2-arithmetic, since we cannot there divide by 2.

PROBLEM 26. We calculate for each of the given equations the *discriminant* $D = b^2 - 4ac$. We obtain $D = 3$ for the first equation, $D = 0$ for the second, and $D = 2$ for the third. We use, for example, the diagram in Problem 21 (Fig. 18) and determine that 3 has no square root in 7-arithmetic, but that 2 has the square roots 3 and 4. Hence, the first equation has no solution, the second exactly one,[1] and the third has two solutions. We find these solutions by formula (3) from Problem 25; for the second equation $x = 2$, and for the third $x_1 = 3$ and $x_2 = 6$.

PROBLEM 27. If α and β are roots of the equation $x^2 + cx + d = 0$, we have as in ordinary algebra

$$x^2 + cx + d = (x - \alpha)(x - \beta).$$

(To be convinced of this it suffices to express α and β in terms of c and d with the aid of the formula for the solution of quadratic equations and to remove the parentheses.) Hence, the number of equations of the required form that have two distinct roots is equal to the number of ways two different numbers α and β can be chosen from the p numbers in p-arithmetic. The number is $\frac{p(p - 1)}{2}$. If the equation $x^2 + cx + d = 0$ has only one root, α, then

$$x^2 + cx + d = (x - \alpha)^2.$$

The number of such equations is equal to p. All other equations (their number is equal to $p^2 - \frac{p(p - 1)}{2} - p = \frac{p(p - 1)}{2}$) have no solution.

---

[1] See solution to Problem 25.

PROBLEM 28. The diagrams for cubing are given below; for 7-arithmetic in Figure 21, for 11-arithmetic in Figure 22, for 13-arithmetic in Figure 23, and for 17-arithmetic in Figure 24.

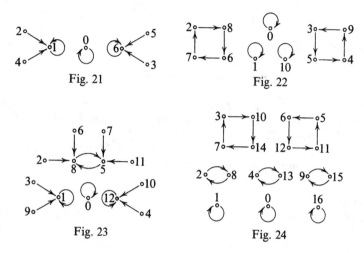

Fig. 21   Fig. 22

Fig. 23   Fig. 24

PROBLEM 29. (a) Let

$$b^3 = 1. \tag{1}$$

By Problem 15,

$$b^{p-1} = 1. \tag{2}$$

Because $p = 3k + 2$, $p - 1 = 3k + 1$ and equality (2) takes the form

$$b^{3k+1} = 1. \tag{3}$$

If equality (1) is raised to the kth power, we obtain

$$b^{3k} = 1, \tag{4}$$

and if we divide (3) by (4), we obtain

$$b = 1. \tag{5}$$

Hence, (5) follows from (1), which was to be proved.

(b) Let $x^3 = a$ and $y^3 = a$. We obtain $\left(\dfrac{x}{y}\right)^3 = 1$ by dividing the first of these equalities by the second. It follows by part (a) that $\dfrac{x}{y} = 1$, i.e., $x = y$.

(c) We construct the following table:

$$0, \ 1, \ 2, \ 3, \ \ldots, \ p - 1,$$
$$0^3, \ 1^3, \ 2^3, \ 3^3, \ \ldots, \ (p - 1)^3.$$

All p numbers in p-arithmetic are in the upper row, and beneath each of them stands its cube. By part (*b*), the lower row consists of p *distinct* numbers, and hence *all* of the numbers in p-arithmetic are contained in it. But this means that the cube root of every number in p-arithmetic can be taken.

PROBLEM 30. It follows from the equation $(x - 1)(x^2 + x + 1) = 0$ that either

$$x - 1 = 0$$

or

$$x^2 + x + 1 = 0.$$

From the first equation we obtain

$$x_1 = 1,$$

and from the second,

$$x_{2,3} = \frac{-1 \pm \sqrt{-3}}{2}. \tag{1}$$

In 103-arithmetic $-3 = 100$ and $\pm\sqrt{-3} = \pm 10$. If we substitute these values for $\pm\sqrt{-3}$ in formula (1), we obtain

$$x_1 = 1, \quad x_2 = 56, \quad x_3 = 46,$$

which are the three cube roots of 1 in 103-arithmetic.

PROBLEM 31. Every value of the cube root of 1 satisfies the equation $x^3 - 1 = 0$. If it is not equal to 1, it satisfies the quadratic equation $x^2 + x + 1 = 0$. Depending on whether or not the square root of -3 can be taken in p-arithmetic, this equation either has two different square roots, which can be calculated from formula (1) of Problem 30, or no square root at all in p-arithmetic. 2-arithmetic and 3-arithmetic are exceptions. Formula (1) (Problem 30) does not make sense in 2-arithmetic. In 3-arithmetic, $-3 = 0$; the equation $x^2 + x + 1 = 0$ has only one root; then 1 has only one cube root, namely 1.

PROBLEM 32. Let $p > 3$, and let there exist an integer a such that $a^2 + 3$ is divisible by p. If the remainder on dividing a by p is equal to b, then in p-arithmetic

$$b^2 + 3 = 0$$

or

$$b^2 = -3.$$

This means that $\sqrt{-3}$ exists in p-arithmetic. It follows from Problem 31 that 1 has three distinct cube roots in p-arithmetic. By Problem 29, p cannot have the form $3k + 2$ and must thus be of the form $3k + 1$.

PROBLEM 33. The equation of the first degree $ax + b = 0$ has only one solution $x = -\dfrac{b}{a}$.

Suppose our theorem has been proved for polynomials of the nth degree. Let us prove that it is then valid also for polynomials of the $(n + 1)$st degree.

We assume the contrary. Let the polynomial of the $(n + 1)$st degree

$$a_0 x^{n+1} + a_1 x^n + \cdots + a_n x + a_{n+1} \ (a_0 \neq 0) \tag{1}$$

have at least $n + 2$ zeros, i.e. $x_1, x_2, \ldots, x_{n+1}, x_{n+2} = 0$. We form the polynomial

$$a_0 (x - x_1)(x - x_2) \cdots (x - x_{n+1}). \tag{2}$$

In this polynomial, the coefficient of x^{n+1} is equal to a_0, and hence in the difference

$$\begin{aligned} &a_0 x^{n+1} + a_1 x^n + \cdots + a_{n+1} \\ &-a_0 (x - x_1)(x - x_2) \cdots (x - x_{n+1}) \end{aligned} \tag{3}$$

the coefficient of x^{n+1} vanishes, and (3) is a polynomial of degree n. But it has the $n + 1$ zeros $x_1, x_2, \ldots, x_{n+1}$. Since by assumption the theorem has been proved for polynomials of the nth degree, all the coefficients of the polynomial (3) are equal to zero. These are, however, the differences of corresponding coefficients of the polynomials (1) and (2); hence, all coefficients of the polynomial (1) are equal to the corresponding coefficients of the polynomial (2); that is, the polynomials (1) and (2) are identical. But then (2) must vanish for $x = x_{n+2}$:

$$a_0 (x_{n+2} - x_1)(x_{n+2} - x_2) \cdots (x_{n+2} - x_{n+1}) = 0,$$

which is impossible, since

$$a_0 \neq 0, \ x_{n+2} - x_1 \neq 0, \ \ldots, \ x_{n+2} - x_{n+1} \neq 0.$$

The theorem is valid for $n = 1$; according to this proof it holds also for $n = 1 + 1 = 2$, for $n = 3$, and in general for all n.

This proof is valid in p-arithmetic, as well as ordinary arithmetic.

PROBLEM 34. Let the remainder on division of k by $p - 1$ be r:

$$k = q(p - 1) + r, \qquad r < p - 1.$$

Then for all x,

$$\begin{aligned} x^{p-1} &= 1, \\ x^{q(p-1)} &= 1, \end{aligned} \tag{1}$$

$$x^k = x^{q(p-1)+r} = 1, \tag{2}$$

from which one obtains by division of (2) by (1): $x^r = 1$, for every x.

If r were greater than 0, the coefficients of the polynomial $x^r - 1$ would not all be equal to zero. However, the number of its zeros $(x = 1, 2, \ldots, p - 1)$ is $p - 1$, which exceeds r, and this contradicts the statement of Problem 33.

PROBLEM 35. We know that

$$\left(a^{\frac{p-1}{2}}\right)^2 = a^{p-1} = 1$$

(Problem 15); then

$$a^{\frac{p-1}{2}} = \pm 1.$$

If $\sqrt{a}$ can be extracted, that is, if a b exists such that $b^2 = a$, then

$$a^{\frac{p-1}{2}} = b^{p-1} = 1.$$

Hence, when $\sqrt{a}$ exists, a is a root of the equation

$$x^{\frac{p-1}{2}} = 1. \tag{1}$$

This equation has no more than $\dfrac{p-1}{2}$ roots (Problem 33). On the other hand, there are exactly $\dfrac{p-1}{2}$ numbers different from zero whose square roots can be taken, and they all satisfy the equation (1). Hence the equation (1) is satisfied *only* by those numbers whose square roots can be taken. Only one possibility remains for the other numbers:

$$a^{\frac{p-1}{2}} = -1.$$

PROBLEM 36. The proof of this theorem is contained in the solution of Problem 33. For the sake of completeness we repeat our train of thought here. The polynomial

$$a_0 x^n + a_1 x^{n-1} + \cdots + a_{n-1} x + a_n$$
$$- a_0 (x - x_1)(x - x_2) \cdots (x - x_n) = 0$$

is a polynomial of the $(n-1)$th degree with n zeros. Hence, all of its coefficients are equal to zero, and the coefficients of the polynomial

$$a_0 x^n + a_1 x^{n-1} + \cdots + a_{n-1} x + a_n$$

coincide with the coefficients of the polynomial

$$a_0(x - x_1)(x - x_2) \cdots (x - x_n).$$

PROBLEM 37. The polynomial $x^{p-1} - 1$ has $p - 1$ zeros, i.e.,

$$1, 2, 3, \ldots, p - 1 = 0.$$

By Problem 36 we have

$$x^{p-1} - 1 = (x - 1)(x - 2)\cdots[x - (p - 1)].$$

For $x = 0$ we obtain

$$
\begin{aligned}
-1 &= (-1)(-2) \cdots [-(p - 1)] \\
&= (-1)^{p-1}\, 1 \cdot 2 \cdot \cdots \cdot (p - 1) \\
&= 1 \cdot 2 \cdot \cdots \cdot (p - 1).
\end{aligned}
$$

PROBLEM 38. *(a)*

```
            197
189⁹) 37233
     −1323
      3591
     −1701
       189
      −189
         0
```
$$37233 \div 189 = 197$$

(b) We divide dividend and divisor by 8:

$$36408 \div 328 = 4551 \div 41$$

```
          111
41¹) 4551
    −41
     451
    −41
      41
    −41
       0
```
$$36408 \div 328 = 111$$

(c) We divide dividend and divisor by 2:

$$851 \div 74 = 425.5 \div 37$$

```
           11.5
37³) 425.5
    −18.5
     407
     −37
       37
     −37
        0
```
$$851 \div 74 = 11.5$$

PROBLEM 39. $1:3 = (6)7$, $1:7 = (285714)3$, $1:9 = (8)9$ (the period is given in parentheses).

PROBLEM 40. (*a*) We denote the *n*th number of the table by x_n. We have

$$x_1 = 5, \quad x_2 = 5^2, \quad x_3 = x_2{}^2 = 5^{2^2}, \quad \ldots, \quad x_n = 5^{2^{n-1}}.$$

We factor the difference $x_{n+1} - x_n$:

$$\begin{aligned}
x_{n+1} - x_n &= 5^{2^n} - 5^{2^{n-1}} = 5^{2^{n-1}}(5^{2^{n-1}} - 1) \\
&= 5^{2^{n-1}}(5^{2^{n-2}} + 1)(5^{2^{n-3}} + 1)(5^{2^{n-4}} + 1) \ldots \\
&\qquad\qquad \ldots (5^2 + 1)(5 + 1)(5 - 1).
\end{aligned}$$

All the factors, with the exception of the first, are even numbers. Since *n* such factors appear, the product is divisible by 2^n. On the other hand, the first factor is divisible by 5^n, since $2^{n-1} \geq n$ for $n \geq 1$. Hence, $x_{n+1} - x_n$ is divisible by 10^n or, what is the same thing, the last *n* digits of the numbers x_{n+1} and x_n coincide.

We now remove the part of the table lying above the enclosed principal diagonal. By what has been already proved, every column of the remaining part of the table contains only one digit. From this it is evident that the last *n* digits of the number x_n coincide with the last *n* digits of the diagonal number $d = \ldots 90625$, or, in other words, $d - x_n = a_n 10^n$, where a_n is some number with infinitely many digits. From the last equality it follows that

$$d = x_n + a_n 10^n,$$
$$d^2 = x_n{}^2 + 2x_n a_n 10^n + a_n{}^2 10^{2n}.$$

By applying the equalities

$$x_n{}^2 = x_{n+1}$$

and

$$d = x_{n+1} + a_{n+1} 10^{n+1},$$

we obtain

$$\begin{aligned}
d^2 - d &= (x_n{}^2 + 2x_n a_n 10^n + a_n{}^2 10^{2n}) - (x_{n+1} + a_{n+1} 10^{n+1}) \\
&= 10^n(2x_n a_n + a_n{}^2 10^n - a_{n+1} 10).
\end{aligned}$$

Hence, the last *n* digits of the numbers d^2 and d coincide. Since *n* is perfectly arbitrary, *all* digits of d^2 and d coincide, that is, $d^2 = d$.

(*b*) We now prove that $e^2 = e$. We have

$$e^2 = (1 - d)^2 = 1 - 2d + d^2 = 1 - 2d + d = 1 - d = e.$$

(*c*) Finally,

$$de = d(1 - d) = d - d^2 = 0.$$

Thus, in the arithmetic of numbers with infinitely many digits, multiplying two numbers different from zero can yield zero. In this respect the arithmetic of numbers with infinitely many digits differs from ordinary arithmetic and resembles 10-arithmetic.

PROBLEM 41. We assume the contrary and consider the product xde. The equation $xd = 1$ implies $xde = 1 \cdot e = e$. The equation $de = 0$ implies $xde = x \cdot 0 = 0$. We obtain a contradiction, since $e \neq 0$.

PROBLEM 42. A number x that is equal to its own square must be a whole number. In fact, if x has k decimal places, where the last is different from zero, x^2 has $2k$ decimal places, the last of which again is different from zero. Hence, for $x^2 = x$ we have $2k = k$, that is, $k = 0$. Furthermore, a number x that is equal to its own square must end in one of the digits 0, 1, 5, or 6; otherwise x and x^2 would end in different digits. Let

$$x^2 = x \qquad \text{and} \qquad y = 1 - x.$$

Then $y^2 = y$, for

$$y^2 = (1 - x)^2 = 1 - 2x + x^2 = 1 - 2x + x = 1 - x = y.$$

If x ends in one of the digits 0 or 5, then y ends in one of the digits 1 or 6, and conversely. Hence, it suffices to find all of the numbers that are equal to their own square and end in one of the digits 0 and 5. Let a and b be two different numbers that both end in either zero or 5. Then the sum $a + b$ ends in zero. Since

$$a^2 - b^2 = (a - b)(a + b),$$

the difference $a^2 - b^2$ ends in more zeros than the difference $a - b$. Hence, $a^2 - b^2 \neq a - b$, which means that either $a^2 \neq a$ or $b^2 \neq b$. In this way we have proved that there is at most one number that is equal to its own square and ends with a zero, and there is also at most one that ends with the digit 5. Such numbers must therefore be 0 or $d = \ldots 90625$.

PROBLEM 43. Let $a = \ldots a_3 a_2 a_1$. The last digit of the product is determined by the product of the last digits of the factors, and hence a^{p-1} has for its last digit the numbers a_1^{p-1} (in the sense of p-arithmetic). But in p-arithmetic $a_1^{p-1} = 1$ (Problem 15). Consequently, the last digit of the number $a^{p-1} - 1$ is zero.

PROBLEM 44. We prove this by induction. The case $k = 1$ has been dealt with in the previous exercise. Suppose the theorem has been proved for $k = l$. We prove it for $k = l + 1$. To do this, we denote $a^{p^{l-1}(p-1)}$ by y and obtain

$$a^{p^l(p-1)} - 1 = y^p - 1^p = (y - 1)(y^{p-1} + y^{p-2} + \cdots + y + 1).$$

Since by the induction hypothesis $y - 1$ ends in l zeros, it suffices to show that

$$y^{p-1} + y^{p-2} + \cdots + 1$$

ends in zero. The last digit of y is 1; hence, the last digits of all the numbers y^2, y^3, ... y^{p-1} are also 1's. From this it follows that the last digit of the number

$$y^{p-1} + y^{p-2} + \cdots + y + 1$$
$$= \underbrace{\cdots 1 + \cdots 1 + \cdots + \cdots 1 + \cdots 1}_{p \text{ times}}$$

is a zero.

Remark. If the result of this exercise is applied to numbers with a finite number of digits, we obtain the following theorem of ordinary arithmetic. If a whole number a is not divisible by a prime number p, then $a^{p^{k-1}(p-1)} - 1$ is divisible by p^k. This theorem can be generalized in the following way.

EULER'S THEOREM. *If a and m are relatively prime, $a^{\phi(m)} - 1$ is divisible by m.*

Here $\phi(m)$ is the number of numbers relatively prime to m which lie between 1 and m. For example, $\phi(10) = 4$. (The numbers relatively prime to 10 between 1 and 10 are 1, 3, 7, 9.) We leave the proof of the relations

$$\phi(p) = p - 1, \ \phi(p^k) = p^{k-1}(p - 1)$$

to the reader.

Euler's theorem holds not only for numbers having a finite number of digits, but also for numbers having an infinite number of digits.

PROBLEM 45. The general form of the nth term of a geometric sequence is $b_n = b_0 q^n$. Hence,

$$b_{n+p^{k-1}(p-1)} = b_0 q^{n+p^{k-1}(p-1)} = b_n q^{p^{k-1}(p-1)}.$$

Now it follows from Problem 44 that the last k digits of the number $q^{p^{k-1}(p-1)}$ are 000...001. But this means that the last k digits of the numbers b_n and $b_n q^{p^{k-1}(p-1)} = b_{n+p^{k-1}(p-1)}$ coincide.

PROBLEM 46. This is solved exactly like Problem 22, but replacing the expression "p-arithmetic" by the expression "arithmetic of p-adic numbers."

PROBLEM 47. This problem is solved exactly like Problem 25, but replacing the expression "p-arithmetic" by the expression "arithmetic of p-adic numbers."

PROBLEM 48. If a number has k significant digits after the decimal point, its square has $2k$ significant digits after the decimal point.

PROBLEM 49. Let the last digit of a be a_1, and the last digit of $b = \sqrt{a}$ be b_1. Then $b_1^2 = a_1$ in p-arithmetic.

PROBLEM 50. We have

$$
\begin{array}{r}
201 \\
\times 201 \\
\hline
201 \\
000 \\
1102 \\
\hline
111101,
\end{array}
$$

which is equal to 112,101 to an accuracy of 3 digits.

PROBLEM 51. We denote the expression $a_{n-1}u_0 + a_nu_1$, by the letter u_n'. The sequence u_0', u_1', u_2', ..., u_n', ... is a Fibonacci sequence. For

$$u_n'{}_{-2} + u_n'{}_{-1} = (a_{n-3}u_0 + a_{n-2}u_1) + (a_{n-2}u_0 + a_{n-1}u_1)$$
$$= (a_{n-3} + a_{n-2})\,u_0 + (a_{n-2} + a_{n-1})\,u_1 = a_{n-1}u_0 + a_nu_1 = u_n'.$$

We further note that

$$u_0' = a_{-1}u_0 + a_0u_1 = u_0, \quad u_1' = a_0u_0 + a_1u_1 = u_1$$

and, hence, the sequences $u_0, u_1, u_2, \ldots, u_n, \ldots$ and $u_0', u_1', u_2', \ldots, u_n', \ldots$ have the first two terms in common. Since the two sequences are Fibonacci sequences, they are fully identical; that is, for any n

$$u_n = u_n' = a_{n-1}u_0 + a_nu_1.$$

PROBLEM 52. It suffices to apply the formula of Problem 51 to the Fibonacci sequence

$$a_{m-1}, a_m, a_{m+1}, a_{m+2}, \ldots, a_{m+(n-1)}, \ldots,$$

in which $u_n = a_{m+(n-1)}$.

PROBLEM 53. If one sets $m = n$ in the formula of the previous exercise, one obtains

$$a_{2n-1} = a_{n-1}{}^2 + a_n{}^2.$$

PROBLEM 54. Let

$$u_{n-1}u_{n+1} - u_n{}^2 = d_n.$$

We then obtain

$$d_{n+1} = u_nu_{n+2} - u_{n+1}{}^2$$
$$= u_n\,(u_n + u_{n+1}) - u_{n+1}{}^2$$
$$= u_n{}^2 + u_nu_{n+1} - u_{n+1}{}^2$$
$$= u_n{}^2 - u_{n+1}\,(-u_n + u_{n+1})$$
$$= u_n{}^2 - u_{n-1}u_{n+1} = -d_n.$$

From this it follows that

$$d_n = -d_{n-1}$$
$$= (-1)^2 d_{n-2}$$
$$= (-1)^3 d_{n-3}$$
$$\cdots$$
$$= (-1)^{n-1} d_1$$
$$= (-1)^{n-1} (u_0 u_2 - u_1^2). \tag{1}$$

Since

$$a_0 a_2 - a_1^2 = -1,$$
$$a_{n-1} a_{n+1} - a_n^2 = (-1)^n.$$

PROBLEM 55. We note that

$$u_{n-1}u_{n+2} - u_n u_{n+1} = u_{n-1}u_n + u_{n-1}u_{n+1} - u_n u_{n+1}$$
$$= u_{n-1}u_{n+1} - u_n(u_{n+1} - u_{n-1})$$
$$= u_{n-1}u_{n+1} - u_n^2$$
$$= (-1)^{n-1}(u_0 u_2 - u_1^2); \tag{1}$$

hence,

$$|u_{n-1}u_{n+2} - u_n u_{n+1}| = |u_0 u_2 - u_1^2|.$$

For the sequence F^0 we have

$$a_{n-1}a_{n+2} - a_n a_{n+1} = (-1)^{n-1}(a_0 a_2 - a_1^2)$$
$$= (-1)^{n-1}(0 \cdot 1 - 1) = (-1)^n.$$

PROBLEM 56. The final digits of the numbers of the Fibonacci sequence F^0 themselves form a Fibonacci sequence in 10-arithmetic (namely the sequence F_{10}^0).

We write out the terms of this sequence:

$$0, 1, 1, 2, 3, 5, 8, 3, 1, 4, 5, 9, 4, 3, 7,$$
$$0, 7, 7, 4, 1, 5, 6, 1, 7, 8, 5, 3, 8, 1, 9,$$
$$0, 9, 9, 8, 7, 5, 2, 7, 9, 6, 5, 1, 6, 7, 3,$$
$$0, 3, 3, 6, 9, 5, 4, 9, 3, 2, 5, 7, 2, 9, 1,$$
$$0, 1, 1, 2, 3, 5, \ldots$$

We note that $c_{60} = 0$ and $c_{61} = 1$. Hence, if we strike out the first 60 terms of the sequence $(c_0, c_1, c_2, \ldots, c_{59})$, there remains a Fibonacci sequence that begins with the terms 0 and 1, that is, the sequence F_{10}^0 once more. Thus, $c_{62} = c_2$, $c_{63} = c_3$, $\ldots$, $c_{60+k} = c_k$, $\ldots$, and F_{10}^0 is periodic, the length of its period being 60.

PROBLEM 57. Let an F_m-sequence $v_0, v_1, v_2, \ldots, v_n, \ldots$ be given. We prove that for a certain r the equalities $v_r = v_0$, $v_{r+1} = v_1$ hold, from which the periodicity of the sequence $v_0, v_1, v_2, \ldots, v_n, \ldots$ can be determined, just as in the previous exercise for $F_{10}{}^0$ (the length of the period being a divisor of r).

In m-arithmetic there are in all m elements. From these no more than m^2 different pairs can be formed. Therefore two equal pairs must occur among the $m^2 + 1$ pairs $(v_0, v_1), (v_1, v_2), (v_2, v_3), \ldots, (v_{m^2}, v_{m^2+1})$. Let two such be the kth and the lth pairs $(k < l \leq m^2 + 1)$; that is, let

$$v_k = v_l, \tag{1}$$

$$v_{k-1} = v_{l-1}. \tag{2}$$

If we subtract equality (2) from equality (1), we obtain

$$v_{k-2} = v_{l-2}. \tag{3}$$

If (3) is subtracted from (2), we find

$$v_{k-3} = v_{l-3}.$$

If we continue subtracting, we finally obtain the equalities

$$v_1 = v_{l-k+1},$$
$$v_0 = v_{l-k}.$$

If we substitute r for $l - k$ in these equalities, we obtain

$$v_1 = v_{r+1},$$
$$v_0 = v_r.$$

Since $l \leq m^2 + 1$ and $k \geq 1$, then $r \leq m^2$.

PROBLEM 58. When the numbers of the sequence F^0 are divided by m, the remainders form the sequence $F_m{}^0$. By Problem 57, this sequence is periodic; the first term, zero, repeats itself infinitely often.

PROBLEM 59. We display the sequence $\widetilde{F}_{11}{}^0$ as an example (Fig. 25).

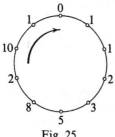

Fig. 25

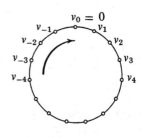

Fig. 26

PROBLEM 60. We denote the element of the sequence that is equal to zero by v_0 and number the other terms as shown in Figure 26. We ascertain that the following equalities hold:

$$v_0 + v_0 = 0,$$
$$v_1 - v_{-1} = 0,$$
$$v_2 + v_{-2} = 0,$$
$$v_3 - v_{-3} = 0,$$
$$v_4 + v_{-4} = 0,$$

. . . .

The first equality holds obviously; the second follows from the fact that

$$v_1 = v_{-1} + v_0 = v_{-1} + 0 = v_{-1}.$$

If we add the first two equalities and take into account that

$$v_0 + v_1 = v_2, \quad v_0 - v_{-1} = v_{-2},$$

we obtain the third equality. By adding the second and the third, we obtain the fourth equality, etc. These equalities can be combined in the following general form:

$$v_n + (-1)^n v_{-n} = 0.$$

PROBLEM 61. The number $y = 0$ is equidistant from the numbers x and z. Hence, by Problem 60 we have either $x + z = 0$ or $x - z = 0$; thus $x = 0$ implies $z = 0$ in either case.

PROBLEM 62. The zeros divide $\breve{F}_m$ into a number of arcs. We choose the smallest of these arcs (if there are several such, we choose one at random) and denote by x and y the elements at the two end points (both are equal to zero). We choose an element z such that the arc $\breve{yz}$ is equal to the arc $\breve{xy}$. Then (see Problem 61) $z = 0$. Now we lay off clockwise the arcs $\breve{zu}$, $\breve{uv}$, etc., each of which is equal to $\breve{xy}$. It is clear that $u = v = \cdots = 0$. When we have traversed the entire circle, we must inevitably have reached either the interior of the arc $\breve{xy}$ or its end x. But no zeros are contained in the interior of the arc $\breve{xy}$. Hence, we can have arrived only at the end point x. Thus, the zeros $x, y, z, u, v, \ldots$ divide the sequence of arcs into equal parts. Since the distance between two zeros is not smaller than the arc $\breve{xy}$, F_m contains no other zeros except those we have constructed, $x, y, z, u, v, \ldots$.

PROBLEM 63. The sequence $F_m{}^0$ consists of the remainders of the terms of F^0 on division by m, and hence the zeros of $F_m{}^0$ correspond to the terms of F^0 divisible by m, and conversely. The sequence $F_m{}^0$ corresponds to the circular sequence $\breve{F}_m{}^0$, which gives its period. Thus, the statement of this exercise follows from Problem 62.

PROBLEM 64. If $\breve{F}_m$ contains zero and contains an odd number of terms, there exists a pair of neighboring terms x, y, that lie at the same distance from the zero (Fig. 27). According to Problem 60 either $x + y = 0$ or $x - y = 0$, and therefore one of the elements of the sequence bordering on the pair x, y is equal to zero. By Problem 61 the other element neighboring the pair x, y is also equal to zero. From this it follows that $x = y$, and since the sequence sought is a sequence without repetitions, it has the form shown in Figure 28.

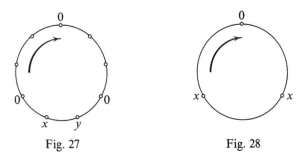

Fig. 27   Fig. 28

PROBLEM 65. Let us suppose that the circular sequence contains at least three different zeros. We choose any three successive zeros and number them 0_1, 0_2, 0_3, clockwise. We denote the elements of the sequence standing just before these elements 0_1, 0_2, 0_3, by u_1, u_2, u_3, and the elements that follow just after the elements 0_1, 0_2, 0_3 by v_1, v_2, v_3 (Fig. 29). Obviously $u_1 = v_1$, $u_2 = v_2$, and $u_3 = v_3$. Since the elements u_1 and v_3 lie at the same distance from 0_2, by Problem 60 either $u_1 + v_3 = 0$ or $u_1 - v_3 = 0$. If $u_1 = v_3$, then $u_1 = v_1 = u_3 = v_3$, which is impossible, since the sequence considered contains no repetitions. Hence, we have the equality $u_1 = -v_3$ and the resulting equality

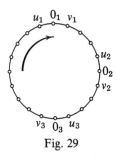

Fig. 29

$$u_3 = -u_1. \tag{1}$$

We now prove that if $\breve{F}_m$ contains exactly three zeros, then it is a sequence with repetitions. In fact, in the case we have considered, the zeros divide the sequence into three equal arcs, and just as we established the equality

$$u_3 = -u_1$$

so we can deduce the equalities

$$u_1 = -u_2, \quad u_2 = -u_3.$$

From these three equalities it follows that $u_1 = u_2 = u_3$, and this means that the sequence considered is a sequence with repetitions.

Suppose now that the sequence contains five different zeros. We prove that in this case also it must contain repetitions. We denote any five successive zeros of the sequence in order by $0_1, 0_2, 0_3, 0_4, 0_5$, and the elements preceding them, clockwise, by u_1, u_2, u_3, u_4, u_5 (Fig. 30).

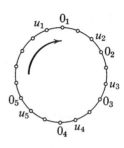

If we apply formula (1) to the different triples of successive zeros that can be formed from the elements $0_1, 0_2, 0_3, 0_4, 0_5$, we obtain the equalities

$$u_3 = -u_1, \quad u_4 = -u_2, \quad u_5 = -u_3,$$

Fig. 30

from which it follows that $u_5 = u_1$; thus, the sequence contains repetitions.

PROBLEM 66. (a) A geometric progression with the initial term b and the common ratio q has the form

$$b, bq, bq^2, bq^3, \ldots, bq^{n-1}, bq^n, bq^{n+1}, \ldots \tag{1}$$

For this sequence to be a Fibonacci sequence, it is necessary and sufficient for

$$bq^{n-1} + bq^n = bq^{n+1}$$

to hold for all n. This relation becomes

$$1 + q = q^2 \quad \text{or} \quad q^2 - q - 1 = 0 \tag{2}$$

after division by bq^{n-1}. We obtain

$$q = \frac{1 \pm \sqrt{5}}{2} \tag{3}$$

as the solution[1] of this quadratic equation for q.

In 11-arithmetic $\sqrt{5} = \pm 4$. Hence,

$$q_1 = \frac{1+4}{2} = 8; \quad q_2 = \frac{1-4}{2} = 4.$$

If one substitutes this value for q in formula (1), one obtains two kinds of geometric progressions that are also Fibonacci sequences:

(1) $b, 8b, 9b, 6b, 4b, 10b, 3b, 2b, 5b, 7b, b, \ldots$

(2) $b, 4b, 5b, 9b, 3b, \quad b, 4b, 5b, 9b, 3b, b, \ldots$

(b) In 7-arithmetic $\sqrt{5}$ does not exist; hence equation (3) has no solution. So there is no geometric progression in 7-arithmetic that is also a Fibonacci sequence.

---

[1] See formula (3) in the solution of Problem 25; we recall that it makes no sense in 2-arithmetic.

PROBLEM 67. In the previous problem we found all sequences in 11-arithmetic that are both geometric progressions and Fibonacci sequences. We established that all these sequences fall into two families. We now choose from the first family a sequence with the initial term x and from the second family a sequence with the initial term y, add corresponding terms, and obtain the sequence

$$x + y, 8x + 4y, 9x + 5y, 6x + 9y, 4x + 3y, 10x + y,$$
$$3x + 4y, 2x + 5y, 5x + 9y, 7x + 3y, x + y, \ldots, \tag{1}$$

which is clearly again a Fibonacci sequence. Let us now choose x and y in such a way that the first term of this sequence is equal to zero and the second term is equal to one. We have to solve the system of equations

$$x + y = 0$$
$$8x + 4y = 1.$$

The solution is

$$x = \frac{1}{4} = 3,$$
$$y = -3 = 8.$$

Thus, $F_{11}{}^0$ is given by:

$$c_0 = 0 = 3 + 8, \quad c_1 = 1 = 3 \cdot 8 + 8 \cdot 4,$$
$$c_2 = 1 = 3 \cdot 8^2 + 8 \cdot 4^2, \ldots, c_n = 3 \cdot 8^n + 8 \cdot 4^n, \ldots$$

PROBLEM 68. We repeat in a more general form the considerations that we have already set forth for a special case in the solution of Problems 66 and 67. The common ratio q of a geometric progression that is at the same time an F-sequence must satisfy the quadratic equation $q^2 - q - 1 = 0$ (see formula (2) in the solution of Problem 66).

(a) If $\sqrt{5}$ does not exist in p-arithmetic, equation (2) of Problem 66 has no solution, from which it follows that the desired Fibonacci sequence cannot exist.

(b) If $\sqrt{5}$ exists, equation (2) of Problem 66 yields two values for q, and each of these values corresponds to a family of geometric sequences that are at the same time Fibonacci sequences:

$$(1) \quad b, bq_1, bq_1{}^2, bq_1{}^3, \ldots, bq_1{}^n, \ldots,$$
$$(2) \quad b, bq_2, bq_2{}^2, bq_2{}^3, \ldots, bq_2{}^n, \ldots.$$

The sequence

$$x + y, xq_1 + yq_2, xq_1{}^2 + yq_2{}^2, xq_1{}^3 + yq_2{}^3, \ldots, xq_1{}^n + yq_2{}^n, \ldots \tag{1}$$

is again a Fibonacci sequence. To represent an arbitrary Fibonacci sequence

$$v_0, v_1, v_2, \ldots, v_n, \ldots \tag{2}$$

in the form (1) it suffices to choose x and y so that the two initial terms of the sequences (1) and (2) coincide. This can be done by solving the system of equations

$$x + y = v_0,$$
$$xq_1 + yq_2 = v_1.$$

We obtain as a solution of this system[1]

$$x = \frac{v_1 - v_0 q_2}{q_1 - q_2}, \qquad y = \frac{v_0 q_1 - v_1}{q_1 - q_2}. \tag{3}$$

Thus, the decomposition of an arbitrary Fibonacci sequence into the sum of two geometric sequences is given by the formula

$$v_n = xq_1^n + yq_2^n, \tag{4}$$

where q_1 and q_2 are the roots of the equation

$$q^2 - q - 1 = 0,$$

and x and y are calculated by formula (3).

PROBLEM 69. According to Problem 15, in p-arithmetic the equation $a^{p-1} = 1$ holds for every $a \neq 0$. In particular, $q_1^{p-1} = 1$ and $q_2^{p-1} = 1$, and it follows from formula (4) in the solution of Problem 68 that

$$v_{p-1} = xq_1^{p-1} + yq_2^{p-1} = x + y = v_0,$$
$$v_p = xq_1^p + yq_2^p = xq_1 + yq_2 = v_1,$$
$$\dots\dots\dots\dots\dots\dots\dots\dots\dots\dots\dots\dots$$
$$v_{k+p-1} = xq_1^{k+p-1} + yq_2^{k+p-1} = xq_1^k + yq_2^k = v_k.$$

PROBLEM 70. We choose some element of the circular sequence $\breve{F}_p$ as a starting point and imagine the circle to be traversed clockwise an infinite number of times. If we write down the elements of this sequence in order, we obtain an infinite periodic sequence $v_0, v_1, v_2, \dots, v_n, \dots$, forming a Fibonacci sequence F_p. According to the preceding exercise we have $v_{p-1} = v_0$ and $v_p = v_1$. Since the original circular sequence is a sequence without repetitions, it does not contain two equal pairs of neighboring elements. Hence the equal pairs (v_0, v_1) and (v_{p-1}, v_p) correspond to one and the same pair of the circular sequence. So the segment $v_0, v_1, v_2, \dots, v_{p-2}$, that is, the first $p - 1$ elements of the sequence F_p, corresponds to a whole number of revolutions about the circular sequence. From this it follows that the number of elements in the circular sequence is a divisor of the number $p - 1$.

[1] Formula (3) makes no sense if $q_1 = q_2$. For this to hold, however, the discriminant of the equation $q^2 - q - 1 = 0$ must be equal to zero. But this discriminant is equal to 5 and can vanish only in 5-arithmetic, and we have excluded this case (see footnote 2, p. 33).

PROBLEM 71. As an example we exhibit the first 15 terms of the sequence $F_{11}{}^0$ and the quotient sequence of $F_{11}{}^0$.

Sequence $F_{11}{}^0$.............	0, 1, 1, 2, 3, 5, 8, 2, 10, 1, 0, 1, 1, 2, 3, 5, ...
Quotient sequence of $F_{11}{}^0$...	∞, 1, 2, 7, 9, 6, 3, 5, 10, 0, ∞, 1, 2, 7, 9, ...

PROBLEM 72. (a) The numbers of a Fibonacci sequence are connected by the relation

$$v_n = v_{n-1} + v_{n-2}.$$

We divide this relation by v_{n-1}: We replace $\dfrac{v_n}{v_{n-1}}$ by t_n and $\dfrac{v_{n-1}}{v_{n-2}}$ by t_{n-1}, and obtain the formula

$$t_n = 1 + \frac{1}{t_{n-1}}. \tag{1}$$

Not more than $p + 1$ different elements appear in the sequence $t_1, t_2, \ldots, t_n, \ldots$: the p numbers of p-arithmetic and the symbol ∞. Hence, equal elements must occur in this sequence. We choose from all pairs of equal elements a pair that has the smallest separation in the sequence considered. Let these be the elements t_k and t_{r+k}. Then any r successive elements of the sequence $t_1, t_2, \ldots, t_n, \ldots$, are all different. For a complete solution of the problem it remains to show that for any value of n, $t_n = t_{r+n}$.

We already know that this equality holds for $n = k$:

$$t_k = t_{r+k}. \tag{2}$$

We now prove the equalities

$$t_{k+1} = t_{r+k+1}, \tag{3}$$

$$t_{k+2} = t_{r+k+2}, \tag{4}$$

$$t_{k+3} = t_{r+k+3}, \tag{5}$$

$$\ldots \ldots \ldots \ldots$$

To do this we use formula (1). By this formula we have

$$t_{k+1} = 1 + \frac{1}{t_k}, \qquad t_{r+k+1} = 1 + \frac{1}{t_{r+k}},$$

and thus equality (3) follows from equality (2). In exactly the same way (4) is deduced from (3), (5) from (4), etc. It remains to ascertain the validity of the equalities

$$t_{k-1} = t_{r+k-1}, \tag{6}$$

$$t_{k-2} = t_{r+k-2}, \tag{7}$$

$$\ldots \ldots \ldots \ldots$$

$$t_1 = t_{r+1}. \tag{8}$$

We solve relation (1) for t_{n-1}:

$$t_{n-1} = \frac{1}{t_n - 1}.$$

With the help of this formula we derive in succession (6) from (2), (7) from (6), etc., just as we derived equations (3), (4), (5), . . . from formula (1).

(*b*) We construct the quotient sequence of $F_p{}^0$. The sequence $F_p{}^0$ begins with the elements 0 and 1. Consequently its quotient sequence begins with the element ∞. According to part (*a*) the symbol ∞ is repeated in the quotient sequence an infinite number of times, and moreover at equal intervals. It is easy to see that these elements of the quotient sequence correspond to the zeros of the sequence $F_p{}^0$. These in turn correspond to the elements of the sequence F^0 that are multiples of p.

PROBLEM 73. First of all we make two observations:

Observation 1. If the quotient sequences

$$t_1, t_2, \ldots, t_n, \ldots \tag{1}$$

and

$$t_1', t_2', \ldots, t_n', \ldots \tag{1'}$$

have the same initial term, they are identical, that is, we have $t_n = t_n'$ for any n. This follows from the formula derived in Problem 72(*a*).

Observation 2. If a finite number of terms are omitted at the beginning of an arbitrary quotient sequence $t_1, t_2, \ldots, t_n, \ldots$, a sequence

$$t_m, t_{m+1}, t_{m+2}, \ldots, t_n, \ldots$$

is obtained that consists of the same elements as the original sequence. In fact, according to Problem 72(*a*) we know that every element of the sequence

$$t_1, t_2, \ldots, t_n, \ldots$$

is repeated an infinite number of times; hence it must occur in the sequence

$$t_m, t_{m+1}, t_{m+2}, \ldots, t_n, \ldots.$$

The converse is clear.

Now suppose we are given two arbitrary quotient sequences (1) and (1'), and we know that a certain element t_m of sequence (1) is equal to a certain element t_n' of sequence (1'). We consider the sequences

$$t_m, t_{m+1}, t_{m+2}, \ldots \tag{2}$$

and

$$t_n', t_{n+1}', t_{n+2}', \ldots. \tag{2'}$$

Sequences (2) and (2') have the same initial term and so are identical by Observation 1. Furthermore, by Observation 2 sequence (2) consists of the same elements as sequence (1), and sequence (2') consists of the same elements as sequence (1'). Hence, sequences (1) and (1') consist of the same elements.

PROBLEM 74. The formula of Problem 51 can be rewritten in the form

$$v_n = c_{n-1}v_0 + c_n v_1$$

for the sequence F_p (see remark on page 29). Using this formula, we derive

$$t_{r+1} = \frac{v_{r+1}}{v_r} = \frac{c_r v_0 + c_{r+1}v_1}{c_{r-1}v_0 + c_r v_1}.$$

We replace c_{r+1} by the equivalent sum $c_{r-1} + c_r$, divide numerator and denominator of the fraction thus transformed by $v_0 c_{r-1}$, and obtain

$$t_{r+1} = \frac{\dfrac{c_r}{c_{r-1}} + \left(1 + \dfrac{c_r}{c_{r-1}}\right)\dfrac{v_1}{v_0}}{1 + \dfrac{c_r}{c_{r-1}}\dfrac{v_1}{v_0}}.$$

Taking into account that $\dfrac{c_r}{c_{r-1}} = \bar{t}_r$ and $\dfrac{v_1}{v_0} = t_1$, we obtain the formula

$$t_{r+1} = \frac{\bar{t}_r + (1 + \bar{t}_r)t_1}{1 + \bar{t}_r t_1}. \tag{1}$$

It follows from equality (1) that:
(1) if $\bar{t}_r = 0$, then $t_{r+1} = t_1$;
(2) if $t_{r+1} = t_1$, then $\bar{t}_r(t_1^2 - t_1 - 1) = 0$, and then either $\bar{t}_r = 0$ or $t_1^2 - t_1 - 1 = 0$.

Now let r denote the length of the period of the sequence $\bar{t}_1, \bar{t}_2, \ldots, \bar{t}_n, \ldots$, and let t_1 not be a root of the equation $x^2 - x - 1 = 0$. We have $\bar{t}_{r+1} = \bar{t}_1 = \infty$, and the formula $\bar{t}_{r+1} = 1 + \dfrac{1}{\bar{t}_r}$ (see Problem 72) implies that $\bar{t}_r = 0$. By Problem 72 again, the elements $\bar{t}_1, \bar{t}_2, \ldots, \bar{t}_r$ are pairwise distinct, and in particular the elements $\bar{t}_1, \bar{t}_2, \ldots, \bar{t}_{r-1}$ are distinct from $\bar{t}_r = 0$. By (1), the equality $\bar{t}_r = 0$ implies the equality $t_{r+1} = t_1$. Moreover, all elements $t_2, t_3, \ldots, t_r$ are different from t_1, since from the equality $t_1 = t_{s+1}$ $(s < r)$ it would follow by (2) that $\bar{t}_s = 0$, which is false. By the same reasoning it is established that the length of the period of the sequence $t_1, t_2, \ldots, t_n, \ldots$ is the same as that of the sequence $\bar{t}_1, \bar{t}_2, \ldots, \bar{t}_n, \ldots$, namely r.

PROBLEM 75. (a) We first consider the case where $\sqrt{5}$ does not exist in p-arithmetic. In this arithmetic the equation $x^2 - x - 1 = 0$ has no solution, and according to Problem 74 all the quotient sequences of any F_p-sequences have the same length of period, r. The elements contained in the period of a quotient sequence are pairwise distinct (by Problem 72(a)). Hence, in every quotient sequence there exist exactly r distinct elements. We denote by R_1 the quotient sequence that corresponds to the sequence $F_p{}^0$ beginning with 0, 1. Its first element is the symbol ∞. The total number of elements that can appear in the quotient sequence is equal to $p + 1$ (the symbol ∞ and the p numbers of p-arithmetic). If all of these elements occur in the sequence R_1, then $r = p + 1$, and the proof is finished. If this is not the case, we choose a number a that is not contained in the sequence R_1, and consider the Fibonacci sequence $1, a, \ldots$. The corresponding quotient sequence (we shall denote it by R_2) begins with the element a. By Problem 73, the sequences R_1 and R_2 have no element in common. If a number b in p-arithmetic is contained in neither of the sequences, we construct the Fibonacci sequence $1, b, \ldots$. The corresponding quotient sequence (we shall denote it by R_3) begins with the element b and therefore has no element in common with either R_1 or R_2. If the sequences R_1, R_2, R_3 still do not exhaust all of the elements $0, 1, \ldots, p - 1, \infty$, we construct a fourth sequence R_4, etc., until we have obtained a system of sequences $R_1, R_2, \ldots,$ R_k that exhausts the whole collection of possible elements. No two of the sequences so constructed have any element in common. Each one of them contains exactly r different elements. Thus, $p + 1 = kr$, and this concludes the proof.

It still remains to consider the second case, in which $\sqrt{5}$ can be extracted in p-arithmetic. In such an arithmetic the equation $x^2 - x - 1 = 0$ has the two solutions $\alpha = \dfrac{1 + \sqrt{5}}{2}$ and $\beta = \dfrac{1 - \sqrt{5}}{2}$. If $p \neq 5$, these solutions are distinct. It follows from the relation $\alpha^2 - \alpha - 1 = 0$ that

$$\alpha^2 = \alpha + 1.$$

On dividing this equality by α we obtain

$$\alpha = 1 + \frac{1}{\alpha}. \tag{1}$$

We now recall the formula in Problem 72(a), which enables us to calculate successively the terms of the quotient sequence, and compare them with formula (1). We see that the quotient sequence that begins with the element α has the form

$$\alpha, \alpha, \alpha, \ldots . \tag{2}$$

Similarly, the quotient sequence that begins with the element β has the form

$$\beta, \beta, \beta, \ldots . \tag{3}$$

According to Problem 74 the periods of all other quotient sequences have the same length r. Following our earlier argument, we construct a system of quotient sequences $R_1, R_2, \ldots, R_k$, that are different from the sequences (2) and (3), have no common elements, and furthermore exhaust all of the elements $0, 1, \ldots, p - 1$, and ∞ (except α and β). We deduce the equality $(p + 1) - 2 = kr$ or $p - 1 = kr$, so that r is a divisor of $p - 1$.

(b) The remainders of the terms of the given sequences on division by p form the p-arithmetic sequence $F_p{}^0$,

$$c_0 = 0,\ c_1 = 1,\ c_2 = 1, \ldots \tag{4}$$

We construct the quotient sequence

$$\bar{t}_1 = \frac{c_1}{c_0} = \infty, \ldots, \bar{t}_n = \frac{c_n}{c_{n-1}}, \ldots \tag{5}$$

of sequence (4).

Assume first that $\sqrt{5}$ does not exist in p-arithmetic. Then by part (a) the length of the period of sequence (5) is a divisor of $p + 1$. Hence, we have $\bar{t}_{p+2} = \bar{t}_{1+(p+1)} = \bar{t}_1 = \infty$, from which it follows that $\frac{c_{p+2}}{c_{p+1}} = \infty$, and hence $c_{p+1} = 0$. This means that a_{p+1} is divisible by p. The case where $\sqrt{5}$ exists in p-arithmetic is handled similarly.

PROBLEM 76. The first nine rows of the triangle in 2-arithmetic (counting the zeroth row) have the form

```
                1
              1   1
            1   0   1
          1   1   1   1
        1   0   0   0   1
      1   1   0   0   1   1
    1   0   1   0   1   0   1
  1   1   1   1   1   1   1   1
1   0   0   0   0   0   0   0   1
```

Here all the terms (except the outside ones) in the second, fourth, and eighth rows are equal to zero. Suppose our assertion to be correct up to the 2^kth row (in this case the $(2^k - 1)$th row consists only of ones). We show that it is also correct for the rows which succeed; that is, that the next row in which all except the end-terms are zero is the 2^{k+1}th row:

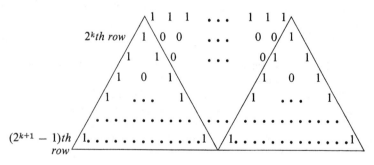

The whole proof is contained in the diagram shown. We urge the reader to consider it carefully before he turns to the detailed proof which follows.

Another dyadic Pascal's triangle identical with the original triangle results from each of the outside terms in the 2^kth row (the new triangles arise from the original triangle by parallel displacement). These new triangles (boxed in the diagram) extend downward until their bases finally come together. This takes place in the row numbered $2^k + h$, which is composed of two equal parts, the hth rows of the two new triangles.

There are $2^k + h + 1$ terms in the $(2^k + h)$th row of the Pascal's triangle. There are $h + 1$ terms in the hth row of each of the new triangles. Therefore,

$$2^k + h + 1 = 2(h + 1),$$
$$h = 2^k - 1.$$

Thus, the $(2^k - 1)$th rows of the new triangles come together. However, the $(2^k - 1)$th row of a new triangle is identical with the $(2^k - 1)$th row of the original triangle, which by our induction hypothesis consists only of ones. Hence, the row numbered

$$2^k + 2^k - 1 = 2^{k+1} - 1$$

of our original triangle likewise consists only of ones. Therefore, the row following this row, the 2^{k+1}th row, consists only of zeros (with the exception of the outside terms).

Furthermore, there is at least one "1" in the interior of any row of the original triangle up to the point where the new triangles come together, that is, up to the 2^{k+1}th row (for example, the outside term in that row of one of the new triangles).

PROBLEM 77. The triangles

$$\triangle_n^0, \triangle_n^1, \ldots, \triangle_n^k, \ldots, \triangle_n^n$$

compose the nth *strip*. Our assertion is verified directly for the first strips. We assume that it is correct up to the nth strip; we prove that it is also true from there on.

The proof is based on the following lemma: Let two groups of numbers be chosen from Pascal's triangle, for example, $a_1, a_2, \ldots, a_r$ and $b_1, b_2, \ldots, b_r$, such that each is an uninterrupted part of some row. Then if the two groups differ from each other only by some factor; that is, if $b_1 = ca_1, b_2 = ca_2, \ldots, b_r = ca_r$, then the triangles $a_1 a_r a$ and $b_1 b_r b$ generated by them (Fig. 31) differ by the same factor. The proof of this lemma is clear and follows directly from the construction of Pascal's triangle.

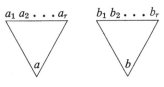

Fig. 31

We consider the triangle $\triangle_n^k$ (Fig. 32). Since by assumption our theorem is valid for the nth strip, $\triangle_n^k = P_n^k \cdot \triangle_0^0$. In particular, one obtains the last row of this triangle from the last row of the triangle $\triangle_0^0$ on multiplying by P_n^k. By virtue of the lemma, the triangle DEF differs from the triangle ABC by the same factor; however, ABC consists of nothing but zeros, and hence DEF also contains nothing but zeros.

From Figure 32 it is clear that if the triangles $\triangle_n^{k-1}$ and $\triangle_n^k$ are generated by the numbers a and b, respectively, and the triangle $\triangle_{n+1}^k$ by the number c, then $c = a + b$, and hence we also have

$$\triangle_{n+1}^k = \triangle_n^{k-1} + \triangle_n^k.$$

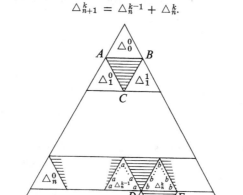

Fig. 32

Furthermore, we have by assumption that

$$a = P_n^{k-1} \cdot 1 = P_n^{k-1}, \quad b = P_n^k \cdot 1 = P_n^k;$$

thus

$$c = P_n^{k-1} + P_n^k = P_{n+1}^k$$

and

$$\triangle_{n+1}^k = P_{n+1}^k \cdot \triangle_0^0.$$

The problem is completely solved.

PROBLEM 78. By virtue of the previous problem, the s^2th row, the first row of the sth strip, contains the upper vertices of the triangles $\triangle_s^0, \triangle_s^1, \ldots, \triangle_s^s$; all other terms of this row are equal to zero. But $\triangle_s^k = P_s^k \cdot \triangle_0^0$, and hence the vertices of the triangles $\triangle_s^k$ are the numbers $P_s^0, P_s^1, \ldots, P_s^{s-1}, P_s^s$. By assumption, $P_s^1 = P_s^2 = \cdots = P_s^{s-1} = 0$.

If one knows that all of the inner terms of the s^2th row are equal to zero, one can prove the assertion in exactly the same way for the s^3th row, and so on.

PROBLEM 79. We carry through the proof by induction:

For $n = 0$, we have $(1 + x)^0 = 1 = C_0^0$.

For $n = 1$, we have $(1 + x)^1 = 1 + x = C_1^0 + C_1^1 x$.

For $n = 2$, we have $(1 + x)^2 = 1 + 2x + x^2 = C_2^0 + C_2^1 x + C_2^2 x^2$.

Suppose this formula has been proved for n. We show that it is also valid for $n + 1$. We have

$$(1 + x)^{n+1} = (1 + x)^n (1 + x)$$
$$= (C_n^0 + C_n^1 x + C_n^2 x^2 + \cdots + C_n^{n-1} x^{n-1} + C_n^n x^n)(1 + x)$$
$$= C_n^0 + C_n^1 x + C_n^2 x^2 + \cdots + C_n^{n-1} x^{n-1} + C_n^n x^n$$
$$+ C_n^0 x + C_n^1 x^2 + \cdots + C_n^{n-2} x^{n-1} + C_n^{n-1} x^n + C_n^n x^{n+1}$$
$$= C_{n+1}^0 + C_{n+1}^1 x + C_{n+1}^2 x^2 + \cdots + C_{n+1}^n x^n + C_{n+1}^{n+1} x^{n+1}$$
$$(C_n^0 = C_n^n = 1 = C_{n+1}^0 = C_{n+1}^{n+1}).$$

If all the numbers in an equality valid in ordinary arithmetic are replaced by their remainders on division by m, we know the result is an equality valid in m-arithmetic. Since P_n^k is the remainder of C_n^k on division by m, the validity of the equality

$$(1 + x)^n = P_n^0 + P_n^1 x + P_n^2 x^2 + \cdots + P_n^n x^n$$

in m-arithmetic follows from the validity of the equality

$$(1 + x)^n = C_n^0 + C_n^1 x + C_n^2 x^2 + \cdots + C_n^n x^n$$

in ordinary arithmetic.

PROBLEM 80. On the basis of the previous problem we have

$$(1 + x)^p = 1 + P_p^1 x + P_p^2 x^2 + \cdots + P_p^{p-1} x^{p-1} + x^p.$$

On the other hand, it follows from Problem 15 that in p-arithmetic

$$x^p = x \quad \text{and}$$

$$(1 + x)^p = 1 + x = 1 + x^p$$

for any x. Hence,

$$(1 + x)^p - (1 + x^p) = 0 = P_p^1 x + P_p^2 x^2 + \cdots + P_p^{p-1} x^{p-1}$$

for every x. The $(p - 1)$th degree polynomial

$$P_p^1 x + P_p^2 x^2 + \cdots + P_p^{p-1} x^{p-1}$$

has p zeros ($x = 0, 1, \ldots, p - 1$); by Problem 33 all of the coefficients of the polynomial are equal to zero; that is,

$$P_p^1 = P_p^2 = \cdots = P_p^{p-1} = 0,$$

which was to be proved.

PROBLEM 81. The diagram for the function defined by $\dfrac{4x + 1}{2x + 3}$ is shown in Figure 33. It consists of two fixed points and three cycles.

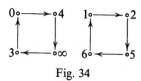

Fig. 33

The diagram for the function defined by $\dfrac{2x + 1}{3x + 2}$ (Fig. 34) consists of two cycles and has no fixed point.

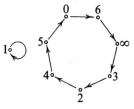

Fig. 34

The diagram for the function defined by $\dfrac{3x - 1}{x + 1}$ (Fig. 35) consists of one fixed point and one cycle.

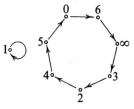

Fig. 35

PROBLEM 82. One obtains the diagram for the function f_{-1} from the diagram of the function f by reversing the directions of all the arrows. Hence, the assertion of the problem follows directly from the fact that in the diagram of any fractional linear function, including f_{-1}, one and only one arrow goes out from each point.

PROBLEM 83. Word for word as in Problem 15(a).

PROBLEM 84. Let $n \neq \infty$ be a fixed point of the fractional linear function defined by $\dfrac{ax + b}{cx + d}$. Then the number n satisfies the equation

$$cx^2 + (d - a)x - b = 0, \tag{1}$$

for we have

$$\frac{an + b}{cn + d} = n, \; an + b = cn^2 + dn,$$
$$cn^2 + (d - a)n - b = 0.$$

We assume $c \neq 0$. Then $f(\infty) = \frac{a}{c} \neq \infty$, and ∞ is not a fixed point. Hence, all fixed points satisfy equation (1). However, the quadratic equation (1) can have no more than two roots for $c \neq 0$ (see Problem 25), whereas in the case under consideration it has at least three roots. Hence, the case $c \neq 0$ is impossible.

So $c = 0$. The point ∞ is transformed into itself: $f(\infty) = \frac{a}{0} = \infty$. The other fixed points (there are at least two) must satisfy equation (1), which now has the form

$$(d - a)x - b = 0. \tag{2}$$

But for $d - a \neq 0$ this equation has the single root $x = \frac{b}{d - a}$. Hence, $d - a = 0$. From this it follows that $b = 0$, and our function has the form $\frac{ax}{a} = x$; that is, it transforms all points into themselves.

PROBLEM 85. We have

$$f(g(x)) = \frac{\dfrac{4x + 3}{6x + 3} + 5}{5\,\dfrac{4x + 3}{6x + 3} + 1} = \frac{6x + 4}{5x + 4}$$

and

$$g(f(x)) = \frac{4\,\dfrac{x + 5}{5x + 1} + 3}{6\,\dfrac{x + 5}{5x + 1} + 3} = \frac{5x + 2}{5} = x + 6.$$

PROBLEM 86. The desired function is defined by $f(x) = 4x$. One can find it in the following way. We write $f(x)$ in the general form

$$f(x) = \frac{ax + b}{cx + d}$$

and substitute for x the successive values 0, 1, 4. We solve the determining equations for a, b, c, d,

$$\frac{a \cdot 0 + b}{c \cdot 0 + d} = 0,$$

$$\frac{a \cdot 1 + b}{c \cdot 1 + d} = 4,$$

$$\frac{a \cdot 4 + b}{c \cdot 4 + d} = 2,$$

to obtain $b = c = 0$, $a = 4d$. So we have $f(x) = \frac{4dx}{d} = 4x$.

PROBLEM 87. Let the desired function be defined by

$$f(x) = \frac{ax + b}{cx + d},$$

and let none of the points x_1, x_2, x_3 equal ∞. By assumption

$$f(x_1) = 0, \text{ and } f(x_3) = \infty;$$

that is,

$$ax_1 + b = 0, \text{ and } cx_3 + d = 0.$$

Hence,

$$b = -ax_1, \qquad d = -cx_3;$$
$$ax + b = ax - ax_1 = a(x - x_1),$$
$$cx + d = cx - cx_3 = c(x - x_3).$$

Thus,

$$\frac{ax + b}{cx + d} = \frac{a(x - x_1)}{c(x - x_3)}.$$

Furthermore,

$$f(x_2) = 1,$$

so that

$$\frac{a(x_2 - x_1)}{c(x_2 - x_3)} = 1,$$

$$\frac{a}{c} = \frac{x_2 - x_3}{x_2 - x_1},$$

from which it follows that

$$f(x) = \frac{x_2 - x_3}{x_2 - x_1} \cdot \frac{x - x_1}{x - x_3}$$

$$= \frac{(x_2 - x_3) x - x_1 (x_2 - x_3)}{(x_2 - x_1) x - x_3 (x_2 - x_1)}.$$

In the case where one of the points x_1, x_2, x_3 is equal to ∞, it is easy to verify that the desired function is defined by

$$f(x) = \frac{x_2 - x_3}{x - x_3} \text{ for } x_1 = \infty,$$

$$f(x) = \frac{x - x_1}{x - x_3} \text{ for } x_2 = \infty,$$

$$f(x) = \frac{x - x_1}{x_2 - x_1} \text{ for } x_3 = \infty.$$

PROBLEM 88. By Problem 87 there exists a function f such that

$$f(y_1) = 0, \qquad f(y_2) = 1, \qquad f(y_3) = \infty.$$

The fractional linear function f_{-1} accomplishes what is desired:

$$f_{-1}(0) = y_1, \qquad f_{-1}(1) = y_2, \qquad f_{-1}(\infty) = y_3.$$

PROBLEM 89. Let f transform x_1, x_2, x_3 into 0, 1, ∞, and ϕ transform 0, 1, ∞ into y_1, y_2, y_3 (such fractional linear functions exist by Problems 87 and 88). We form a new function defined by $\phi(f(x))$. It is easy to see that this function transforms x_1, x_2, x_3 into y_1, y_2, y_3, respectively, and that it is fractional linear.

Suppose that two fractional linear functions g and h satisfy the conditions

$$g(x_1) = y_1, g(x_2) = y_2, g(x_3) = y_3;$$
$$h(x_1) = y_1, h(x_2) = y_2, h(x_3) = y_3.$$

We wish to show that g and h are identical. We have

$$g_{-1}(y_1) = x_1, g_{-1}(y_2) = x_2, g_{-1}(y_3) = x_3;$$
$$g_{-1}(h(x_1)) = x_1, g_{-1}(h(x_2)) = x_2, g_{-1}(h(x_3)) = x_3.$$

The function $g_{-1}(h)$ transforms three distinct points into themselves; by Problem 84, it leaves all points fixed; that is, the equation $g_{-1}(h(n)) = n$ holds for any n. However, if g_{-1} transforms $m = h(n)$ into n, then g transforms the number n into $m = h(n)$; that is, we have $g(n) = h(n)$ for every n, which is what was to be proved.

PROBLEM 90. We choose a triple of distinct points, say 0, 1, ∞. Every fractional linear transformation transforms this triple into another triple of distinct points, y_1, y_2, y_3. Conversely, for every triple of distinct points, y_1, y_2, y_3, one can find exactly one fractional linear function that transforms the triple 0, 1, ∞ into this triple. Hence, there are as many different fractional linear functions in p-arithmetic as there are triples of distinct points in this arithmetic enlarged with the symbol ∞ (triples that differ in the order of the points are to be regarded as distinct). For y_1 we can choose any of the $p + 1$ points, $0, 1, \ldots, p - 1, \infty$, for y_2 any of the remaining p points, and, finally, for y_3, any of the remaining $p - 1$ points. Thus, there are

$$(p + 1)p(p - 1)$$

such combinations, that is,

$$(p + 1)p(p - 1)$$

distinct triples of points and just as many fractional linear functions in p-arithmetic.

172 MATHEMATICAL CONVERSATIONS

PROBLEM 91. The diagram for the function defined by $f(x) = \dfrac{3x - 1}{x + 1}$ is given in Figure 36. The diagrams for the functions f_2, f_3, and f_4, respectively, are shown in Figures 37, 38, and 39.

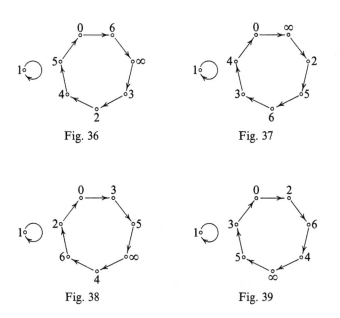

Fig. 36   Fig. 37

Fig. 38   Fig. 39

PROBLEM 92. On the application of function f_k to the point x_0, this point moves k steps on its cycle. Since k is divisible by s, it makes a number of complete revolutions and returns to its starting point. Thus,

$$f_k(x_0) = x_0.$$

Conversely, if the point returns to its starting point after k steps, this means that it has completely traversed its cycle a certain number of times, so that k is divisible by s.

PROBLEM 93. Let n be a point that is transformed into itself by f_k but not by f; i.e., $f_k(n) = n$ and $f(n) \neq n$. If $f_k(n) = n$, all points on the same cycle of f as n are transformed into themselves by f_k. This follows directly from Problem 92. It follows from Problem 84 that f_k transforms *all* points into themselves if the length of this cycle is greater than two. The case in which the length of the cycle considered is equal to two and the function f leaves at least one point fixed is dealt with in just the same way.

It remains to consider the case in which f has a cycle of length two and transforms no point into itself.

We calculate the function f_2 defined by

$$f_2(x) = \frac{(a^2 + bc)x + ab + bd}{(ac + cd)x + bc + d^2}.$$

The points of our cycle are transformed into themselves by f_2. At least one of them is not equal to ∞; this point satisfies the equation (see the solution of Problem 84)

$$(ac + cd)x^2 + (bc + d^2 - a^2 - bc)x - (ab + bd) = 0,$$
$$c(a + d)x^2 + (d - a)(d + a)x - b(a + d) = 0,$$
$$[cx^2 + (d - a)x - b](a + d) = 0. \tag{1}$$

If $a + d \neq 0$, then

$$cx^2 + (d - a)x - b = 0;$$

but this means (see the solution of Problem 84) that x is transformed into itself by f, which contradicts our assumption that f transforms no point into itself. Hence,

$$a + d = 0.$$

But then *all* numbers satisfy equation (1); that is, all points are transformed into themselves by f_2; $f_2(x_0) = x_0$ for any x_0. By Problem 92 all cycles have length 2 (for there are no fixed points, and therefore no cycles of length one).

In particular, the fixed point of the function f_k also lies on a cycle of length two, and according to Problem 92, k is an even number. Since the length of any cycle is a divisor of k, we obtain $f_k(n) = n$ for arbitrary n, which was to be proved.

PROBLEM 94. We exclude all fixed points and consider a cycle of smallest length r among the remaining cycles. The points of this cycle are transformed into themselves by the function f_r. Hence, by Problem 93 the function $f_r(x)$ transforms all points into themselves.

We now choose any cycle consisting of at least two terms; let its length be s. The function f_r transforms each of its points into itself; hence (Problem 92) r is divisible by s. Since $r \leq s$, $r = s$. Thus, the length of any cycle is equal to r.

PROBLEM 95. If we exclude the uninteresting case in which all points are transformed into themselves by the function f (in this case $x_{p+1} = x_p = x_{p-1} = x_0$), then either no point at all, or one, or two points are transformed into themselves by the function f (Problem 84). The other points are divided into cycles of equal length (Problem 94); the length of these cycles is a divisor of one of the numbers $p + 1$, p, $p - 1$, depending on which of the above cases holds true. In the first case $x_{p+1} = x_0$; in the second, $x_p = x_0$;

and in the third, $x_{p-1} = x_0$ (Problem 92). Which of these cases occurs depends on the number of fixed points. If $c \neq 0$, ∞ is not a fixed point, and the number of fixed points corresponds to the number of roots of the equation (Problem 84)

$$cx^2 + (d - a)x - b = 0.$$

By Problem 25 we know that this quadratic equation has no root if the square root of its discriminant

$$(a - d)^2 + 4bc$$

cannot be taken, one root when the discriminant is equal to zero, and two roots when the discriminant is different from zero and its square root *can* be taken.

We leave it to the reader to investigate the case $c = 0$.

Remark. The quotient sequence

$$t_1, t_2, \ldots, t_n, \ldots$$

of the Fibonacci sequence (see p. 34) is a special case of the general sequence

$$x_0, x_1, \ldots, x_k, \ldots$$

Here $t_n = 1 + \dfrac{1}{t_{n-1}}$, $f(t) = 1 + \dfrac{1}{t} = \dfrac{t + 1}{t}$, $a = b = c = 1$, and $d = 0$. The equation for a fixed point is

$$x^2 - x - 1 = 0.$$

Its discriminant is equal to 5. We see that the results of Problems 74 and 75(*a*) are special cases of the general result for an arbitrary fractional linear transformation.

PROBLEM 96.

$$f(x) = \frac{x - 3}{x + 1},$$

$$f_2(x) = \frac{\dfrac{x - 3}{x + 1} - 3}{\dfrac{x - 3}{x + 1} + 1} = \frac{-x - 3}{x - 1},$$

$$f_3(x) = \frac{\dfrac{-x - 3}{x - 1} - 3}{\dfrac{-x - 3}{x - 1} + 1} = x.$$

The function f_3 leaves all points fixed. By Problem 92, the number three is divisible by the length of any cycle of the function f. Hence, the length of a cycle can be only 1 or 3. Now the function f does not transform all points into themselves. For example, $f(\infty) = 1$. Hence, not all cycles have the length 1; that is, there exists at least one cycle of length 3.

PROBLEM 97. (*a*) We consider the function defined by

$$f(x) = \frac{x-3}{x+1}$$

in *p*-arithmetic. We have

$$(a-d)^2 + 4bc = (1-1)^2 + 4(-3) \cdot 1 = 4(-3).$$

Suppose that $\sqrt{-3}$ exists. Then

$$\sqrt{(a-d)^2 + 4bc} = \sqrt{4(-3)} = 2\sqrt{-3}$$

also exists. By Problem 95 the equation $f_{p-1}(x_0) = x_0$ holds for every x_0, and in particular for such x_0 as lie on a cycle of length 3 (as we have seen in Problem 96, such a cycle exists). Hence (Problem 92), $p - 1$ is divisible by 3, $p - 1 = 3k$, $p = 3k + 1$.

Suppose now that $\sqrt{-3}$ does not exist. Then

$$\sqrt{(a-d)^2 + 4bc} = 2\sqrt{-3}$$

cannot be extracted, and $f_{p+1}(x_0) = x_0$ for every x_0. From this it follows that

$$p + 1 = 3l, \quad p = 3l - 1 \neq 3k + 1.$$

(*b*) Let *p* be a prime divisor of the number $a^2 + 3$, and let *b* be the remainder on dividing *a* by *p*. Then in *p*-arithmetic $b^2 + 3 = 0$, $b^2 = -3$; that is, $\sqrt{-3}$ exists in *p*-arithmetic. Therefore, $p = 3k + 1$.

Now let $p = 3k + 1$. Then $\sqrt{-3}$ exists in *p*-arithmetic; that is, one can find a number *a* such that $a^2 + 3 = 0$ in *p*-arithmetic. This means that in ordinary arithmetic $a^2 + 3$ is divisible by *p*.

PROBLEM 98.

$$f(x) = \frac{x-1}{x+1}; \quad f_2(x) = \frac{\dfrac{x-1}{x+1} - 1}{\dfrac{x-1}{x+1} + 1} = -\frac{1}{x};$$

$$f_3(x) = \frac{-\dfrac{1}{x} - 1}{-\dfrac{1}{x} + 1} = \frac{-x-1}{x-1}; \quad f_4(x) = \frac{\dfrac{-x-1}{x-1} - 1}{\dfrac{-x-1}{x-1} + 1} = x.$$

The function f_4 transforms every point into itself. By Problem 92, the lengths of the cycles of *f* must be divisors of the number 4; that is, the cycles must have the lengths 1, 2, or 4. But not all cycles have the length 1, since not all points are transformed into themselves by the function *f*, e.g.,

$$f(0) = -1.$$

If all cycles had the length 1 or 2, the function f_2 would transform all points into themselves. But this is not the case, for

$$f_2(0) = \infty.$$

Hence, there exists at least one cycle of length 4.

PROBLEM 99. (*a*) We consider the function defined by

$$f(x) = \frac{x-1}{x+1}$$

in *p*-arithmetic. Here

$$(a-d)^2 + 4bc = (1-1)^2 + 4(-1)\cdot 1 = 4(-1).$$

Suppose that $\sqrt{-1}$ exists. Then

$$\sqrt{(a-d)^2 + 4bc} = \sqrt{4(-1)} = 2\sqrt{-1}$$

exists. By Problem 95 we have $f_{p-1}(x_0) = x_0$ for every x_0, in particular for such x_0 as lie on a cycle of length 4 (as we have seen in Problem 98, such a cycle exists). Hence (Problem 92), $p - 1$ is divisible by 4, $p - 1 = 4k$, $p = 4k + 1$.

If $\sqrt{-1}$ does not exist, then

$$\sqrt{(a-d)^2 + 4bc} = 2\sqrt{-1}$$

does not exist either, and $f_{p+1}(x_0) = x_0$ for every x_0. From this it follows that

$$p + 1 = 4l, p = 4l - 1 \neq 4k + 1.$$

(*b*) The solution is word for word that of Problem 24(*b*).

PROBLEM 100. The geometric sequence

$$b, bq, bq^2, \ldots, bq^n, \ldots$$

is also a Fibonacci sequence if and only if its common ratio q satisfies the equation $q^2 - q - 1 = 0$ (see the solution of Problem 66). Solving this quadratic equation, we obtain

$$q_1 = \frac{1 + \sqrt{5}}{2}, \quad q_2 = \frac{1 - \sqrt{5}}{2}$$

(see the solution of Problem 47). A solution thus exists if and only if $\sqrt{5}$ exists in *p*-adic arithmetic. We now apply the result formulated on page 27. It follows from this theorem that if *p* is distinct from 2 and 5, the condition that $\sqrt{5}$ exists in the arithmetic of *p*-adic numbers[1] is equivalent to the condition that $\sqrt{5}$ exists in *p*-arithmetic. If $\sqrt{5}$ exists, we obtain two families of *p*-adic *F*-sequences that are at the same time geometric sequences:

$$b, bq_1, bq_1^2, \ldots, bq_1^n, \ldots,$$
$$b, bq_2, bq_2^2, \ldots, bq_2^n, \ldots.$$

PROBLEM 101. The solution follows from Problem 100 (compare with Problem 68).

---

[1] It would perhaps be better to write $\sqrt{\ldots005}$, to avoid confusion with $\sqrt{5}$, where 5 is an element of *p*-arithmetic.

PROBLEM 102. According to the previous problem, one can write the *F*-sequence u_0, u_1, u_2, ..., u_n, ... as the sum of two geometric sequences; hence, the solution follows from Problem 45.

PROBLEM 103. We write the sequence 0, 1, 1, 2, 3, ... in the *p*-adic system. On the basis of the previous problem, the last *k* digits of the terms a_0 and $a_{p^{k-1}(p-1)}$ coincide. But all of the digits of a_0 are zeros ($a_0 = \ldots 000$). Hence, the last *k* digits of $a_{p^{k-1}(p-1)}$ are also zeros. But this means that $a_{p^{k-1}(p-1)}$ is divisible by p^k.

PROBLEM 104. We check this formula for the first rows of the triangle:

$$C_0^0 = \frac{0!}{0!0!} = 1$$

$$C_1^0 = \frac{1!}{0!1!} = 1 \qquad\qquad C_1^1 = \frac{1!}{1!0!} = 1$$

$$C_2^0 = \frac{2!}{0!2!} = 1 \qquad\qquad C_2^1 = \frac{2!}{1!1!} = 2 \qquad\qquad C_2^2 = \frac{2!}{2!0!} = 1$$

We now assume that our formula holds for the *n*th row, and show that it then holds for the $(n + 1)$st row also. Let us write out the *n*th and the $(n + 1)$st rows:

$$C_n^0 \quad C_n^1 \quad \cdots \quad C_n^{k-1} \quad C_n^k \quad \cdots \quad C_n^n$$
$$C_{n+1}^0 \quad C_{n+1}^1 \quad \cdots \quad C_{n+1}^{k-1} \quad C_{n+1}^k \quad C_{n+1}^{k+1} \quad \cdots \quad C_{n+1}^{n+1}$$

We use the relation

$$C_{n+1}^k = C_n^{k-1} + C_n^k.$$

Since, by assumption,

$$C_n^{k-1} = \frac{n!}{(k-1)![n-(k-1)]!} = \frac{n!}{(k-1)!(n-k+1)!},$$
$$C_n^k = \frac{n!}{k!(n-k)!},$$

we have

$$C_{n+1}^k = \frac{n!}{(k-1)!(n-k+1)!} + \frac{n!}{k!(n-k)!}$$
$$= \frac{n!k}{k!(n-k+1)!} + \frac{n!(n-k+1)}{k!(n-k+1)!}$$
$$= \frac{n!(k+n-k+1)}{k!(n-k+1)!}$$
$$= \frac{(n+1)!}{k![(n+1)-k]!}.$$

Our formula is valid for the zeroth, first, and second rows. Therefore it holds also for the third row, the fourth row, ...; that is, it holds for all rows. (From the formula we have proved, it follows that $n!$ is divisible by the product $k!(n-k)!$ for any $k < n$.)

PROBLEM 105. We have

$$
\begin{aligned}
C_{p-k-1}^k &= \frac{(p-k-1)!}{k!(p-2k-1)!} \\
&= \frac{(p-k-1)(p-k-2)\cdots(p-2k)(p-2k-1)\cdots 1}{k!(p-2k-1)(p-2k-2)\cdots 1} \\
&= \frac{(p-k-1)(p-k-2)\cdots(p-2k)}{k!} \\
&= \frac{[p-(k+1)][p-(k+2)]\cdots[p-2k]}{k!}.
\end{aligned}
$$

Now $k!$ is not divisible by p, since $k \leq \dfrac{p-1}{2} < p$. So, $k!$ is not equal to zero in p-arithmetic. Thus, we obtain the equality

$$
\begin{aligned}
P_{p-k-1}^k &= \frac{[-(k+1)][-(k+2)]\cdots[-2k]}{k!} \\
&= \frac{(-1)^k(k+1)(k+2)\cdots 2k}{k!} \\
&= (-1)^k\frac{(2k)!}{k!k!} \\
&= (-1)^k P_{2k}^k.
\end{aligned}
$$

PROBLEM 106. We have

$$
P_{2n}^n = \frac{(2n)!}{n!n!}.
$$

We note that

$$
\begin{aligned}
(2n)! &= 1\cdot 2\cdot 3\cdots\cdots 2n = 1\cdot 3\cdot 5\cdots\cdots(2n-1)\cdot 2\cdot 4\cdot 6\cdots\cdots 2n \\
&= 1\cdot 3\cdot 5\cdots\cdots(2n-1)\cdot(2\cdot 1)\cdot(2\cdot 2)\cdots\cdots(2\cdot n) \\
&= 1\cdot 3\cdot 5\cdots\cdots(2n-1)\cdot 2^n\cdot 1\cdot 2\cdot 3\cdots\cdots n \\
&= 2^n\cdot n!\cdot 1\cdot 3\cdot 5\cdots\cdots(2n-1).
\end{aligned}
$$

Hence,

$$
P_{2n}^n = 2^n\frac{1\cdot 3\cdot 5\cdots\cdots(2n-1)}{n!}.
$$

At the same time,

$$1 \cdot 3 \cdot 5 \cdots (2n - 1) = (-1)^n(-1)(-3) \cdots [-(2n - 1)]$$
$$= (-1)^n(p - 1)(p - 3) \cdots [p - (2n - 1)]$$
$$= (-1)^n 2^n \left(\frac{p - 1}{2}\right)\left(\frac{p - 3}{2}\right) \cdots \left(\frac{p - 2n + 1}{2}\right)$$
$$= (-1)^n 2^n \left(\frac{p - 1}{2}\right)\left(\frac{p - 1}{2} - 1\right) \cdots \left(\frac{p - 1}{2} - n + 1\right)$$
$$= (-1)^n 2^n \frac{\left(\frac{p - 1}{2}\right)!}{\left(\frac{p - 1}{2} - n\right)!},$$

from which it follows that

$$P_{2n}^n = (-1)^n 4^n \frac{\left(\frac{p - 1}{2}\right)!}{n!\left(\frac{p - 1}{2} - n\right)!} = (-4)^n P_{\frac{p-1}{2}}^n .$$

PROBLEM 107. By Problems 106 and 79,

$$S = 1 + 2q + \cdots + P_{2n}^n q^n + \cdots + P_{p-1}^{\frac{p-1}{2}} q^{\frac{p-1}{2}}$$
$$= 1 - 4P_{\frac{p-1}{2}}^1 q + \cdots + (-4)^n P_{\frac{p-1}{2}}^n q^n + \cdots + (-4)^{\frac{p-1}{2}} P_{\frac{p-1}{2}}^{\frac{p-1}{2}} q^{\frac{p-1}{2}}$$
$$= 1 + P_{\frac{p-1}{2}}^1(-4q) + \cdots + P_{\frac{p-1}{2}}^n(-4q)^n + \cdots + P_{\frac{p-1}{2}}^{\frac{p-1}{2}}(-4q)^{\frac{p-1}{2}}$$
$$= (1 - 4q)^{\frac{p-1}{2}} .$$

If $q \neq \frac{1}{4}$, then $1 - 4q \neq 0$ and $S^2 = (1 - 4q)^{p-1} = 1$ (Problem 15), from which it follows that $S = \pm 1$. If, however, $q = \frac{1}{4}$, then $1 - 4q = 0$, and $S = 0$.

PROBLEM 108. It is clear from Figure 11 (p. 51), that $b_{n-1} + b_n = b_{n+1}$. Thus, the numbers b_n form a Fibonacci sequence. But $b_1 = 1$, $b_2 = 1$; that is, the initial terms of the Fibonacci sequences

$$a_1 = 1, a_2 = 1, a_3 = 2, \ldots ,$$
$$b_1, \quad b_2, \quad b_3, \ldots ,$$

are identical. Consequently, these sequences are identical in their entirety: $b_n = a_n$.

PROBLEM 109. We know that a_p is the sum of the numbers that lie on the pth diagonal. By passing to p-arithmetic, we obtain

$$c_p = P_{p-1}^0 + P_{p-2}^1 + \cdots + P_{p-1-k}^k + \cdots + P_{p-1-\frac{p-1}{2}}^{\frac{p-1}{2}}$$

$$= P_{p-1}^0 + P_{p-2}^1 + \cdots + P_{p-1-k}^k + \cdots + P_{\frac{p-1}{2}}^{\frac{p-1}{2}}.$$

(The sum breaks off here; $P_{\frac{p-1}{2}}^{\frac{p-1}{2}}$ is the last term of the $\dfrac{p-1}{2}$th row.) But

$$P_{p-1-k}^k = (-1)^k P_{2k}^k \text{ (Problem 105).}$$

Hence,

$$c_p = P_0^0 - P_2^1 + \cdots + (-1)^k P_{2k}^k + \cdots + (-1)^{\frac{p-1}{2}} P_{p-1}^{\frac{p-1}{2}}.$$

If we apply the formula found in the solution of Problem 107 and take into account that in this case $q = -1$, we obtain

$$c_p = [1-4(-1)]^{\frac{p-1}{2}} = 5^{\frac{p-1}{2}}.$$

PROBLEM 110. The proof is carried out by induction. For $n = 2$ we obviously have

$$d_0 \qquad d_1$$
$$d_0^{(1)}$$
$$d_0^{(1)} = d_0 + d_1 = C_1^0 d_0 + C_1^1 d_1.$$

Assume our theorem proved for n numbers. We prove it for $n + 1$ numbers. To do this we consider the table

$$d_0 \quad d_1 \quad d_2 \quad \cdots \quad d_{n-2} \quad d_{n-1} \quad d_n$$
$$d_0^{(1)} \quad d_1^{(1)} \qquad\qquad d_{n-2}^{(1)} \quad d_{n-1}^{(1)}$$
$$d_0^{(2)} \qquad\qquad\qquad d_{n-2}^{(2)}$$
$$\cdots \quad d_0^{(n-1)} \quad d_1^{(n-1)} \quad \cdots$$
$$d_0^{(n)}$$

It follows from the induction hypothesis that

$$d_0^{(n-1)} = C_{n-1}^0 d_0 + C_{n-1}^1 d_1 + \cdots + C_{n-1}^{n-1} d_{n-1},$$
$$d_1^{(n-1)} = C_{n-1}^0 d_1 + \cdots + C_{n-1}^{n-2} d_{n-1} + C_{n-1}^{n-1} d_n.$$

By adding these equalities we obtain

$$d_0^{(n)} = d_0^{(n-1)} + d_1^{(n-1)}$$
$$= C_n^0 d_0 + C_n^1 d_1 + \cdots + C_n^{n-1} d_{n-1} + C_n^n d_n.$$

PROBLEM 111. We prove that

$$(C_n^0)^2 + (C_n^1)^2 + \cdots + (C_n^n)^2 = C_{2n}^n.$$

We apply the result of Problem 110 to the table

$$
\begin{array}{ccccc}
C_n^0 & C_n^1 & C_n^2 & \cdots & C_n^n \\
C_{n+1}^1 & C_{n+1}^2 & \cdots & C_{n+1}^n \\
& & \ddots & & \\
& & C_{2n}^n & &
\end{array}
$$

We have

$$d_0 = C_n^0, \ d_1 = C_n^1, \ \ldots, \ d_n = C_n^n, \ d_0^{(n)} = C_{2n}^n.$$

Hence,

$$C_{2n}^n = C_n^0 C_n^0 + C_n^1 C_n^1 + \cdots + C_n^n C_n^n$$
$$= (C_n^0)^2 + (C_n^1)^2 + \cdots + (C_n^n)^2.$$

PROBLEM 112. We write out the table

$$
\begin{array}{cccccccc}
v_k & v_{k+1} & v_{k+2} & v_{k+3} & \cdots & v_{k+p-2} & v_{k+p-1} & v_{k+p} \\
& v_{k+2} & v_{k+3} & v_{k+4} & \cdots & v_{k+p} & v_{k+p+1} & \\
& & v_{k+4} & v_{k+5} & \cdots & v_{k+p+2} & & \\
& & & \ddots & & \ddots & & \\
& & & & v_{k+2p} & & &
\end{array}
$$

Here under every pair of neighboring numbers there stands their sum.
Now we apply the result of Problem 110. We have

$$d_0 = v_k, \quad d_1 = v_{k+1}, \quad \ldots, \quad d_p = v_{k+p};$$
$$d_0^{(p)} = v_{k+2p}.$$

In p-arithmetic

$$v_{k+2p} = P_p^0 v_k + \cdots + P_p^p v_{k+p}.$$

But

$$P_p^0 = P_p^p = 1 \text{ and } P_p^1 = P_p^2 = \cdots = P_p^{p-1} = 0 \qquad \text{(Problem 80)}.$$

Hence,

$$v_{k+2p} = v_k + v_{k+p}.$$

PROBLEM 113. The terms of the F_p-sequence whose indices are divisible by p form the sequence $v_0, v_p, v_{2p}, \ldots, v_{np}, \ldots$. If we substitute, in the formula of Problem 112, $k = (n-1)p$, we obtain

$$v_{(n-1)p} + v_{np} = v_{(n+1)p}.$$

PROBLEM 114. We must multiply together m parentheses:

$$\underbrace{(a + b)(a + b)\cdots(a + b).}_{m \text{ times}}$$

If the parentheses are removed and the terms combined, a^m obviously has the coefficient 1. To obtain $a^{m-1}b$ one must multiply b in one of the parentheses by a in all of the others. Hence, $a^{m-1}b$ has the coefficient m. All other terms will contain a power of b not smaller than 2.

PROBLEM 115. The assertion obviously holds for $l = 1$. Now suppose it is valid for $l = s$; we prove that it is then also valid for $l = s + 1$. We set $n = k$ and $m = ks$ in Problem 52. We then obtain

$$a_{k(s+1)-1} = a_{k+ks-1} = a_{k-1}a_{ks-1} + a_k a_{ks}.$$

Since both a_k and $a_0 = 0$ are divisible by d, then by Problem 63 a_{ks} also is divisible by d. Hence, the product $a_k a_{ks}$ is divisible by d^2 and

$$a_{k(s+1)-1} = a_{k-1}a_{ks-1} + xd^2, \tag{1}$$

where x is a whole number. By assumption,

$$a_{ks-1} = a_{k-1}{}^s + yd^2. \tag{2}$$

By substituting (2) in (1) we obtain

$$a_{k(s+1)-1} = a_{k-1}a_{k-1}{}^s + d^2(ya_{k-1} + x) = a_{k-1}{}^{s+1} + zd^2,$$

so that $a_{k(s+1)-1} - a_{k-1}{}^{s+1}$ is divisible by d^2. One proves in exactly the same way that $a_{kl+1} - a_{k+1}{}^l$ is divisible by d^2.

PROBLEM 116. By assumption, $a_k = xm^n$, where x is a whole number. From this it follows that

$$a_{k+1} = a_{k-1} + a_k = a_{k-1} + xm^n \text{ and}$$
$$a_{k+1}{}^m = (a_{k-1} + xm^n)^m.$$

By Problem 114

$$(a_{k-1} + xm^n)^m = a_{k-1}{}^m + ma_{k-1}{}^{m-1}xm^n + x^2m^{2n}S$$
$$= a_{k-1}{}^m + m^{n+1}(a_{k-1}{}^{m-1}x + x^2m^{n-1}S).$$

Hence, $a_{k+1}{}^m - a_{k-1}{}^m$ is divisible by m^{n+1}.

PROBLEM 117. By Problem 115 (setting $d = m^n$)

$$a_{km+1} = a_{k+1}{}^m + xm^{2n},$$
$$a_{km-1} = a_{k-1}{}^m + ym^{2n}.$$

According to the previous problem, $a_{k+1}{}^m - a_{k-1}{}^m$ is divisible by m^{n+1}. Consequently the difference $a_{km+1} - a_{km-1} = a_{km}$ is also divisible by m^{n+1}.

PROBLEM 118. The solution follows from the previous problem. If a_k is divisible by m, then a_{km} is divisible by m^2, and then, also, a_{km^2} is divisible by $m^3, \ldots$; $a_{km^{n-1}}$ is divisible by m^n. By Problem 63 all of the terms $a_{km^{n-1}s}$ are divisible by m^n if $a_{km^{n-1}}$ is divisible by m^n.

PROBLEM 119. Suppose $b \neq d$. Since

$$a - c = (d - b)\sqrt{5},$$

the integer $a - c$ must be equal to the irrational number $(d - b)\sqrt{5}$. This contradiction shows that

$$d - b = 0,$$

whence

$$a - c = 0.$$

Thus,

$$b = d, \quad a = c.$$

PROBLEM 120. We have

$$(a + b\sqrt{5})(c + d\sqrt{5}) = ac + bc\sqrt{5} + ad\sqrt{5} + 5bd$$
$$= (ac + 5bd) + (ad + bc)\sqrt{5}.$$

The second statement of this problem follows directly from this formula.

PROBLEM 121. By using the solution of Problem 120, we obtain

$$m + n\sqrt{5} = (ac + 5bd) + (ad + bc)\sqrt{5}.$$

By Problem 119 we have $m = ac + 5bd$ and $n = ad + bc$. Hence,

$$m - n\sqrt{5} = (ac + 5bd) - (ad + bc)\sqrt{5}$$
$$= [ac + 5(-b)(-d)] + [a(-d) + (-b)c]\sqrt{5}$$
$$= (a - b\sqrt{5})(c - d\sqrt{5}).$$

PROBLEM 122. (a) If $a^2 - 5b^2 = 1$, then also $a^2 - 5(-b)^2 = 1$. So

$$a + (-b)\sqrt{5} = a - b\sqrt{5}$$

is likewise a solution of equation (1).

$$(b) \quad \frac{1}{a + b\sqrt{5}} = \frac{a - b\sqrt{5}}{(a + b\sqrt{5})(a - b\sqrt{5})}$$
$$= \frac{a - b\sqrt{5}}{a^2 - 5b^2}$$
$$= \frac{a - b\sqrt{5}}{1}$$
$$= a - b\sqrt{5}.$$

PROBLEM 123. (*a*) Since by assumption

$$m + n\sqrt{5} = (a + b\sqrt{5})(c + d\sqrt{5}),$$

by Problem 121

$$m - n\sqrt{5} = (a - b\sqrt{5})(c - d\sqrt{5}).$$

Hence,

$$
\begin{aligned}
m^2 - 5n^2 &= (m + n\sqrt{5})(m - n\sqrt{5}) \\
&= (a + b\sqrt{5})(c + d\sqrt{5})(a - b\sqrt{5})(c - d\sqrt{5}) \\
&= (a + b\sqrt{5})(a - b\sqrt{5})(c + d\sqrt{5})(c - d\sqrt{5}) \\
&= (a^2 - 5b^2)(c^2 - 5d^2) \\
&= 1 \cdot 1 \\
&= 1.
\end{aligned}
$$

$$
(b) \quad
\begin{aligned}
\frac{a + b\sqrt{5}}{c + d\sqrt{5}} &= (a + b\sqrt{5})\frac{1}{c + d\sqrt{5}} \\
&= (a + b\sqrt{5})(c - d\sqrt{5}).
\end{aligned}
$$

On the basis of Problem 122(*a*), $c - d\sqrt{5}$ is a solution of equation (1). The product

$$(a + b\sqrt{5})(c - d\sqrt{5})$$

can be represented in the form $p + q\sqrt{5}$ (Problem 120) and is a solution of equation (1) (Problem 123(*a*)).

PROBLEM 124. $9^2 - 5 \cdot 4^2 = 1$, and hence

$$9 + 4\sqrt{5}$$

is a solution of equation (1). By Problem 123(*a*)

$$
\begin{aligned}
p_2 + q_2\sqrt{5} &= (9 + 4\sqrt{5})(9 + 4\sqrt{5}) \\
&= (9 + 4\sqrt{5})^2
\end{aligned}
$$

is also a solution of equation (1). Similarly,

$$
\begin{aligned}
p_3 + q_3\sqrt{5} &= (p_2 + q_2\sqrt{5})(9 + 4\sqrt{5}) \\
&= (9 + 4\sqrt{5})^3
\end{aligned}
$$

is a solution, and in general

$$p_n + q_n\sqrt{5} = (9 + 4\sqrt{5})^n$$

are solutions for any positive integer n. All of these solutions are distinct, since for $m \neq n$

$$(9 + 4\sqrt{5})^m \neq (9 + 4\sqrt{5})^n.$$

PROBLEM 125. The case $a \geq c$ and $b \geq d$ is excluded, since otherwise we would have

$$a + b\sqrt{5} \geq c + d\sqrt{5}.$$

Hence, at least one of the inequalities $a < c$ or $b < d$ holds. We will show that the validity of one of these inequalities implies the validity of the other. Let $a < c$. Then

$$a^2 < c^2$$

and

$$b^2 = \frac{a^2 - 1}{5} < \frac{c^2 - 1}{5} = d^2,$$

from which it follows that $b < d$, since $b \geq 0$ and $d \geq 0$. Similarly, if $b < d$, then

$$a^2 = 5b^2 + 1 < 5d^2 + 1 = c^2,$$

and, thus, $a < c$.

PROBLEM 126. (*a*) Suppose $a < 0$. Then at least one of the numbers

$$a + b\sqrt{5} \qquad \text{and} \qquad a - b\sqrt{5}$$

must be negative (if only because their sum is equal to $2a$, which is negative). At the same time the product

$$(a + b\sqrt{5})(a - b\sqrt{5}) = a^2 - 5b^2 = 1$$

is positive. Hence, both are negative, which contradicts

$$a + b\sqrt{5} > 0.$$

(*b*) As we have shown, $a \geq 0$. Suppose $b \leq 0$. Then

$$-b \geq 0 \geq b$$

and

$$a - b\sqrt{5} \geq a + b\sqrt{5} > 1.$$

From this it follows that

$$1 = a^2 - 5b^2 = (a + b\sqrt{5})(a - b\sqrt{5}) > 1 \cdot 1 = 1,$$

which is a contradiction.

PROBLEM 127. Suppose there exists such a solution. It follows from Problem 126 that $a \geq 0$ and $b > 0$. Hence, by Problem 125, $a < 9$ and $b < 4$. Thus, b can assume only the values 1, 2, 3. One can easily verify that none of the numbers $1 + 5b^2$, where $b = 1, 2, 3$, is a perfect square; hence, the number $a + \sqrt{5} \cdot b$ with $b = 1, 2, 3$, cannot be a solution of equation (1).

PROBLEM 128. For $n = 0$ we have

$$(9 + 4\sqrt{5})^0 = 1 = 1 + 0\sqrt{5}$$

and for $n = 1$, we have

$$(9 + 4\sqrt{5})^1 = 9 + 4\sqrt{5},$$

where $1 + 0\sqrt{5}$ and $9 + 4\sqrt{5}$ are solutions of equation (1). For $n \geq 2$ the formula

$$(9 + 4\sqrt{5})^n$$

yields the product of solutions of equation (1), that is, new solutions $p + q\sqrt{5}$, where $p \geq 0$ and $q \geq 0$ (Problem 120).

We now show that our formula yields the totality of integral solutions $p + q\sqrt{5}$ such that $p \geq 0$ and $q \geq 0$. To do this we assume the contrary. Let $p + q\sqrt{5}$ with $p \geq 0$ and $q \geq 0$ be a solution that is not identical with any term of the sequence

$$(9 + 4\sqrt{5})^0, (9 + 4\sqrt{5})^1, (9 + 4\sqrt{5})^2, \ldots, (9 + 4\sqrt{5})^n, \ldots.$$

The case where

$$p + q\sqrt{5} < (9 + 4\sqrt{5})^0 = 1 + 0 \cdot \sqrt{5}$$

is excluded (for then we would have $q < 0$, by Problem 125). Since the terms of our sequence increase without limit, $p + q\sqrt{5}$ must lie between some pair of these terms; that is, there exists an index $n \geq 0$ such that

$$(9 + 4\sqrt{5})^n < p + q\sqrt{5} < (9 + 4\sqrt{5})^{n+1}.$$

We divide this inequality by $(9 + 4\sqrt{5})^n$ and obtain

$$1 < \frac{p + q\sqrt{5}}{(9 + 4\sqrt{5})^n} < 9 + 4\sqrt{5}.$$

Now $\dfrac{p + q\sqrt{5}}{(9 + 4\sqrt{5})^n}$, being the quotient of two solutions of equation (1), is again a solution of equation (1). In this way we obtain a solution of equation (1) that lies between 1 and $9 + 4\sqrt{5}$, and this contradicts Problem 127.

PROBLEM 129. If $a + b\sqrt{5}$ is a solution of equation (1), then

$$a - b\sqrt{5}, \quad -a + b\sqrt{5}, \quad -a - b\sqrt{5}$$

are obviously also solutions of equation (1). By Problem 128 one of these four numbers can be represented in the form $(9 + 4\sqrt{5})^n$. Then the other three can be represented in the forms

$$(9 + 4\sqrt{5})^{-n} = \frac{1}{(9 + 4\sqrt{5})^n}, \quad -(9 + 4\sqrt{5})^n, \quad -(9 + 4\sqrt{5})^{-n}.$$

PROBLEM 130. Suppose
$$a + x = a$$
$$a + y = a.$$

Choose a' so that
$$a + a' = x,$$

and a'' so that
$$a + a'' = y.$$

This is possible by (3).

Then
$$a + x = a,$$
so $a'' + (a + x) = a'' + a,$
so $(a'' + a) + x = a'' + a$ (by (2)),
so $(a + a'') + x = a + a''$ (by (1)),
and $y + x = y.$

By considering similarly the sum
$$a' + (a + y) = a' + a$$
we find that $x + y = x.$

But $x + y = y + x$ (by (1)).

So $x = y.$

PROBLEM 131. Given b, choose z so that
$$b + z = b$$
and b_1, so that $b + b_1 = a.$

Then
$$a = b + b_1 = b_1 + b = b_1 + (b + z)$$
$$= (b_1 + b) + z = (b + b_1) + z = a + z.$$

So $a = a + z.$

It follows from Problem 130 that $z = x$.

PROBLEM 132. Given c in the field, let d be any element of the field.

Then $cd = c(d + 0) = cd + c0$ (by (6)).

So $c0 = 0$ (by the result of Problem 131).

SECTION THREE
Random Walks

Introduction

There is a well-known board game called *Circus*. Two players take turns rolling a die, and each in turn moves his piece forward on a square board which is divided into 100 numbered squares. The piece is moved as many squares forward as the spots on the die indicate. At the beginning of the game, both pieces are placed on the first square. The winner is the one who first reaches the square numbered 100. There is one further rule: If a piece reaches a square with a red number, it moves to another square (either forward or backward) whose number is blue; this move is specified by the "circus act" marked on the board. Obviously, in this game the motion of the pieces depends not on the skill of the players, but rather on chance (assuming the die is rolled "fairly"). The motion of these pieces is one simple example of a *random walk*.

We give another example of a random walk. Two friends live in a city whose map is shown in Figure 1. They leave their house,

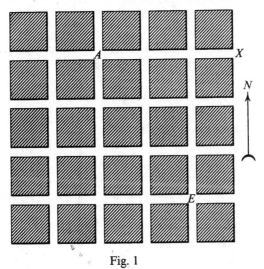

Fig. 1

which stands at the intersection *A*, and set out to go for a walk. However, they disagree as to the route they are to take; they agree that at each intersection, beginning with A, they will toss two coins and proceed north, east, south, or west, depending on which of the four possible tosses (heads-heads, heads-tails, tails-heads, tails-tails) comes up. Thus, they toss the coins and begin their walk at the intersection A; when they reach the following intersection, they again toss the coins and choose their further path according to the result. If they come to the edge of town (for example, to the point X), they will turn around and come back.

Cases similar to the examples just given (but much more complicated) are encountered in nature. Brownian motion will serve as an example: "If a light powder is suspended in water, and if a drop of this water, together with the particles contained therein, is placed under the microscope, one can observe that the particles appear to be alive, because they are in continuous zigzag motion." Here, the random path (Fig. 2) has an essentially more complicated character than in the previous example: in the first place, the particles can change the direction of their motion at any moment, while in the example of the walk in the city, this was possible only at intersections; in the second place, the previous example permitted only four possible directions of motion, whereas the particles can move in any direction.

Fig. 2

In this section, we shall consider a few simple examples of random walks. We return to the board game described above and investigate the duration of the game. To solve this problem we consider a less complicated board (Fig. 3), one that has only 25 squares, and add the rule that a piece landing on square 24 must automatically return to the starting point. There are no other rules for moving from square to square, and the squares are all colored with the same color. How long does it take to play a complete game on this board? It is possible for a player to reach square 25 as early as the fourth move; this occurs if he rolls four successive sixes. On the other hand, if each player's piece repeatedly lands on square

21	22	23	24	25
20	19	18	17	16
11	12	13	14	15
10	9	8	7	6
1	2	3	4	5

Fig. 3

24, the game will not have ended after 1,000 rolls of the die. It can even happen that the game never ends at all; this occurs, for example, if both players roll the following sequence:

6, 6, 6, 5, 6, 6, 6, 5, 6, 6, 6, 5, 6, 6, 6, 5, 6, 6, 6, 5, ...

In this case, no bound can be given for the duration of the game. But try playing one game after another. You will be convinced that the game does end and, in fact, rather quickly. Just what is the trouble here?

We have already determined that it is impossible to give a definite bound for the duration of the game. The situation is basically altered, however, when an absolutely certain answer is not required. To clarify this last remark, let us consider some more examples.

EXAMPLE 1. Suppose we have an urn which contains 1,000 balls; suppose, also, that one of the balls is black, the others white. A ball is taken from this urn at random. Can it happen that the black ball is chosen?

Answer. It is certainly possible, but it is not at all likely.

EXAMPLE 2. Suppose a man who does not know Russian starts to type on a Russian typewriter. Is it possible that he thereby writes Pushkin's story, *The Captain's Daughter?*

Answer. Obviously, this is not absolutely impossible. However, it hardly appears to be a real possibility to anyone. (Indeed, the essays of two school children may also coincide word for word. But, as a rule, one quite correctly regards this as evidence that one of the essays was copied.)

EXAMPLE 3. According to the kinetic theory of gases, air is composed of a great number of molecules that are moving randomly. There is almost no interaction between the individual molecules; hence, the position of one molecule in space does not influence the positions of the other molecules. We imagine the space in the room we are in to be divided into an upper and a lower half. It is not impossible (this follows from the kinetic theory of gases) for all of the molecules of air in the room to go into the upper half, suffocating everyone in the room. This conclusion appears far-fetched to us. But does this mean that the kinetic theory is false?

Answer. It is to be hoped that the reader does not draw this conclusion, in view of the examples discussed above. The phenomenon that we have just presented is indeed not absolutely impossible; one can, however, view it as practically impossible. The basis for this conclusion is even more apparent in this example than in the previous example: there are incomparably more molecules in a room than there are letters in the story *The Captain's Daughter.*

Thus, we may quite correctly regard a highly unlikely occurrence as impossible for all practical purposes. Moreover, there are different degrees of unlikelihood. The unlikelihood of the occurrence of the event in the second example is much greater than that in the first, and the unlikelihood in the third example is much greater than that in the second.

We return again to our problem, which we now reformulate as follows: What *limit* can be estimated for the length of the game, so that a game whose length exceeds the estimated limit is practically impossible in the same degree as are the events of our examples?

How one solves this and similar problems will become clear to the reader after working through this book. We shall give only the answer here. If we estimate that the game ends after at most 200 throws, an error will be about as unlikely as the event in the first example. If the estimate is increased to 30 million, an error will be about as unlikely as the event in the second example. If the estimate is raised to 10^{25}, an error will be as unlikely as the event in the third example.

1. Probability

1. FUNDAMENTAL PROPERTIES OF PROBABILITY

First we shall learn how to calculate probabilities. Consider two urns with 100 balls in each. In the first urn, one ball is white, the other 99 black. In the second urn, 10 balls are white and 90 black. From which urn is one more likely to draw a white ball?[1] The reader will answer without hesitation: from the second. If we ask how many times greater this probability is, the reader will certainly answer that it is ten times as great. Suppose, now, that we have a third urn, in which all 100 balls are white. As before, we conclude that a white ball can be drawn from the third urn with 100 times the probability as from the first urn. The ball taken from the third urn will surely be a white ball. If we define the probability of this last, inevitable event to be the number 1, it follows from what has been said that the probability of drawing a white ball from the first urn is equal to $\frac{1}{100}$, and the probability of drawing a white ball from the second urn is $\frac{10}{100}$.

We consider the general case, in which the urn contains n balls, of which m are white. From the same considerations as above, we can conclude that the probability of a white ball being drawn from the urn is $\frac{m}{n}$. The urn problem is extraordinarily useful because many problems can be reduced to this form.

EXAMPLE 1. What is the probability that heads will come up on the toss of a coin? We consider an urn with one white and one black ball. Let the white ball correspond to heads, the black to tails.

Answer. Obviously, the probability that heads will come up is equal to the probability that one draws a white ball from our urn, and this equals $\frac{1}{2}$.

---

[1] Here it is assumed that the balls in the urns are completely uniform and well mixed and that one does not look when drawing; then, one has the same probability of drawing any ball.

EXAMPLE 2. What is the probability that one rolls a five with a die? The problem may be thought of as an urn with six balls, one of which is white. What is the probability that the white ball will be drawn?

Answer. $\frac{1}{6}$.

EXAMPLE 3. A domino is drawn at random from a box of dominos. What is the probability that there is a six on one end of this domino? We consider an urn with 28 balls, of which 7 (those corresponding to the 7 dominos that have a six on one of their halves) are white.

Answer. The probability that a domino of the desired kind is drawn is equal to the probability that a white ball is drawn from the urn, that is, $\frac{7}{28}$.

EXAMPLE 4. There are 5 red, 7 blue, and 13 black balls in an urn. What is the probability that either a red or a blue ball is drawn? There are 12 white and 13 black balls in a second urn. What is the probability that a white ball is drawn?

Answer. The probability that either a red or a blue ball is drawn from the first urn is equal to the probability that a white ball is drawn from the second urn, that is, $\frac{12}{25}$.

In general, a trial may have n equally probable outcomes, of which m yield a desired event A, the others yielding an undesired event. Each such trial is equivalent to drawing a ball from an urn containing n balls, of which m are white and the rest black. The occurrence of the event A has exactly the same probability as the drawing of a white ball from the urn, that is,

$$\frac{m}{n}.$$

DEFINITION. *The* probability *of an event A is equal to the number of possible favorable outcomes divided by the total number of possible outcomes.*

We denote the *probability* of the event A by

$$\mathbf{P}(A).$$

We now formulate the following properties of probability:

Property 1. If the event *A implies* the event *B*, that is, if each occurrence of the event *A* is followed by an occurrence of event *B* (or, the event *B* always occurs when the event *A* does), then

$$\mathbf{P}(A) \leq \mathbf{P}(B).$$

Property 2. If the events *A* and *B* are *mutually exclusive* (that is, it is impossible that both *A* and *B* occur), then

$$\mathbf{P}(A + B) = \mathbf{P}(A) + \mathbf{P}(B), \tag{1}$$

where *A + B* is understood to mean the event that consists of the occurrence of either *A* or *B*.

Property 3. If the events *A* and *B* are *exact opposites of each other,* that is, if the occurrence of *A* is the same as the nonoccurrence of *B*, then

$$\mathbf{P}(A) + \mathbf{P}(B) = 1. \tag{2}$$

Property 4. If the event *E* is *certain,* that is, if *E* must occur, then

$$\mathbf{P}(E) = 1.$$

Property 5. If the event *O* is *impossible,* that is, if *O* cannot occur, then

$$\mathbf{P}(O) = 0.$$

It is easy to obtain these properties from a consideraton of the urn problem, and the reader may derive them for himself.[1] We shall only clarify a few definitions: an example of *mutually exclusive* events is the drawing of a blue ball and the drawing of a red ball (when only one ball can be drawn from the urn); an example of *opposite results* is the tossing of a coin, where either heads or tails must come up. The drawing of a white ball from an urn that contains only white balls is a certain event; the drawing of a black ball from this urn would then be an impossible event.

Although many problems can be reduced to the urn problem, there are many (and among them the most interesting) that cannot be reduced to this problem. However, Properties 1–5 of probability are always true.

---

[1] We have already derived a special case of Property 2 in Example 4 on page 196. Here we restate the Example: *A* is the event of a red ball being drawn; *B* is the event of a blue ball being drawn; and *A + B* becomes the event of either a red or a blue ball being drawn.

2. CONDITIONAL PROBABILITY

We now wish to become acquainted with so-called *conditional probability;* first, a few examples.

At recess, students of the first and second grades gather in the playground to play. Eleven pupils of the first grade take part, 8 boys and 3 girls, together with 6 pupils of the second grade, 2 boys and 4 girls. It is decided by lot who is to begin.[1] What is the probability that the lot falls on a first-grade student? To calculate this probability, the number of students of the first grade must be divided by the number of all participants. The result equals $\frac{11}{17}$.

We now assume that we know that the game will be started by a boy, and ask what influence this has on the probability of interest to us. We want to know what the probability is now for a member of the first grade to start the game. All 10 boys that participate in the game are equally likely to begin. Of these 10 boys, 8 are pupils in the first grade. Hence, the probability that a member of the first grade begins is, in this case, equal to $\frac{8}{10}$. We see that the probability has changed.

We have obtained the *conditional probability* that a member of the first grade begins, *on condition* that the game is begun by a boy.

DEFINITION. *The* conditional probability *of an event B given an event A is the probability that the event B will occur, if it is known that a previous event A is certain to occur.* It is denoted by

$$\mathbf{P}(B|A).$$

Problem 1. In the example above calculate the conditional probability that a member of the first grade begins, on condition that the game is begun by a girl.

In our example, let B denote "the game is begun by a member of the first grade" and A denote "a boy begins." As we calculated, $\mathbf{P}(B) = \frac{11}{17}$ and $\mathbf{P}(B|A) = \frac{8}{10}$. Hence, in this case $\mathbf{P}(B|A) \neq \mathbf{P}(B)$. The occurrence of the event A thus has a significant influence on the probability of the event B.

We add yet another property of probability to the five already enumerated in section 1.

---

[1] The drawing must guarantee that the probability of any participant beginning the game is the same. Among the possible forms of the drawing, that one is best which achieves this equality of probability most completely.

Property 6.

$$\mathbf{P}(AB) = \mathbf{P}(A)\,\mathbf{P}(B|A). \tag{3}$$

By AB, we are to understand the event in which both A and B occur.

We shall now indicate how one can derive this property. We shall verify Property 6 by means of our previous example of the students in the playground. The person who begins the game is determined by lot. As before, let A be the event "A boy begins" and B the event "A member of the first grade begins." Then, AB is the event "A boy of the first grade begins." Of the 17 possible outcomes, the event AB can occur in 8 ways (8 boys are pupils in the first grade); the event A can occur in 10 ways. Hence, $\mathbf{P}(AB) = \frac{8}{17}$ and $\mathbf{P}(A) = \frac{10}{17}$. We have already found that $\mathbf{P}(B|A) = \frac{8}{10}$; thus we have

$$\frac{8}{17} = \frac{10}{17}\cdot\frac{8}{10},$$

i.e.,

$$\mathbf{P}(AB) = \mathbf{P}(A)\,\mathbf{P}(B|A).$$

If A and B are independent events, neither the occurrence nor nonoccurrence of the event A has any effect on the probability of the event B; hence, the conditional probability $\mathbf{P}(B|A)$ of the event B on the condition A is equal to the unconditional probability $\mathbf{P}(B)$:

$$\mathbf{P}(B|A) = \mathbf{P}(B).$$

In this case, formula (3) takes the form

$$\mathbf{P}(AB) = \mathbf{P}(A)\,\mathbf{P}(B),$$

and we obtain:

Property 6a. If A and B are independent events, then

$$\mathbf{P}(AB) = \mathbf{P}(A)\,\mathbf{P}(B). \tag{4}$$

EXAMPLE. Successive flips of a coin are independent events. Hence, the probability of obtaining heads twice in a row is $\mathbf{P}$("heads come up on the first toss" and "heads come up on the second toss") $=$ $\mathbf{P}$("heads come up on the first toss") $\cdot \mathbf{P}$("heads come up on the second toss") $= \frac{1}{2}\cdot\frac{1}{2} = \frac{1}{4}$.

Formulas (1) and (4), the formulas for addition and multiplication of probabilities, can easily be generalized to the case of more than two mutually exclusive or independent events. Let n events $A_1, A_2, \ldots, A_n$, any two of which are mutually exclusive, be given. Then ·

$$\mathbf{P}(A_1 + A_2 + A_3 + \cdots + A_n)$$
$$= \mathbf{P}(A_1) + \mathbf{P}(A_2) + \mathbf{P}(A_3) + \cdots + \mathbf{P}(A_n). \quad (5)$$

We prove this formula for $n = 3$. The event A_3 is mutually exclusive of both A_1 and A_2; from this it clearly follows that A_3 and $A_1 + A_2$ are mutually exclusive. Then, by Property 2, we have the following formula

$$\mathbf{P}(A_1 + A_2 + A_3) = \mathbf{P}(A_1 + A_2) + \mathbf{P}(A_3). \quad (5')$$

But A_1 and A_2 are also mutually exclusive, so that

$$\mathbf{P}(A_1 + A_2) = \mathbf{P}(A_1) + \mathbf{P}(A_2). \quad (5'')$$

Formula (5) follows, for the case $n = 3$, from (5') and (5''). Formula (5) can be proved analogously for arbitrary n.

If A_1 and A_2 are independent, and if A_1A_2 and A_3 are also independent, then

$$\mathbf{P}(A_1A_2A_3) = \mathbf{P}(A_1A_2)\mathbf{P}(A_3) = \mathbf{P}(A_1)\mathbf{P}(A_2)\mathbf{P}(A_3).$$

More generally, if A_1 and A_2 are independent, and, also, each of the pairs of events A_1A_2 and A_3, $A_1A_2A_3$ and A_4, $\ldots$, $A_1A_2A_3 \cdots A_{n-1}$ and A_n are independent, then

$$\mathbf{P}(A_1A_2A_3 \cdots A_n) = \mathbf{P}(A_1)\mathbf{P}(A_2)\mathbf{P}(A_3) \cdots \mathbf{P}(A_n). \quad (6)$$

For example, let A_k be the event that the kth toss of a coin is heads. Then, the probability that the coin comes up heads n times is given by formula (6).

$$\mathbf{P}(A_1A_2 \cdots A_n) = \mathbf{P}(A_1)\mathbf{P}(A_2) \cdots \mathbf{P}(A_n) = \underbrace{\tfrac{1}{2} \cdot \tfrac{1}{2} \cdot \cdots \cdot \tfrac{1}{2}}_{n \text{ times}} = \frac{1}{2^n}.$$

This same number gives the probability that the coin comes up tails n times and, in general, that on n tosses any previously specified sequence of heads and tails comes up.

Problem 2. What is the probability that no six will come up on six rolls of a die?

Problem 3. We understand by p the probability that the target is hit on one shot. Calculate the probability that in n shots, one hits the target.

We mentioned in the introduction that events of small probability can, with good grounds, be regarded as practically impossible, and that there are varying degrees of unlikelihood. We shall make this more precise.

If one is to apply the methods of probability theory to the study of a phenomenon in nature, he must each time choose an arbitrarily small number ε, *the permissible probability of a deviation* (*an error*). If we have predicted a course of events by arguments based on probability theory, we must admit the possibility of error in our prediction and demand only that the probability of this error is not greater than ε. We start with the assumption that all events whose probability is smaller than ε are to be regarded as *practically impossible,* and that all events whose probability is greater than $1 - \varepsilon$ can be assumed to be *practically certain.*

DEFINITION. *The number ε is the* permissible degree (*or the* magnitude) of uncertainty *and the number* $1 - \varepsilon$ *is the* required degree (*or* magnitude) of certainty.

Obviously, the value of ε must be chosen for each individual problem according to the practical requirements on the correctness of the conclusions. Frequently used values of ε are 0.01, 0.005, 0.001, and 0.0001.

Let us clarify our definition and discussion by considering the example of the number of shots that are necessary for a single hit on a target (see Problem 3 above). Suppose that for each shot the probability of a hit is 0.2. How many shots must be made to hit the target once? It is clear that the number of shots cannot be given with absolute certainty. For it can happen that the target is hit on the first shot; at the same time, one cannot exclude the possibility that after 100 or 200 shots none have yet hit the target. Hence, we shall not seek after absolute certainty, but rather introduce a *permissible degree of uncertainty* ε. It appears entirely acceptable, for example, to set the value ε equal to 0.001. The statement that the target will have been hit after n shots is false with probability $(1 - 0.2)^n = (0.8)^n$.

We now choose n so that

$$(0.8)^n < 0.001.$$

The smallest value of n that satisfies this inequality is 31. (It is easy to calculate this with the aid of a table of logarithms.) Hence, the statement that the target will be hit at least once after 31 shots is false with a probability that does not exceed the permissible bound 0.001. Hence, under our requirements for the degree of certainty, we can say that it is practically certain that the target will be hit after 31 shots.

The number n varies with the required degree of certainty ε. The values of n for different degrees of certainty are compiled in the following table:

ε	n
0.01	21
0.005	24
0.001	31
0.0001	42

EXAMPLE. Calculate the probability that no six comes up on infinitely many rolls of a die.

SOLUTION. We first calculate the probability of the event B_n that no six has come up after n rolls. In the solution of Problem 2, page 10, we found $\mathbf{P}(B_6) = \left(\dfrac{5}{6}\right)^6$. Analogously, we find (by formula (6)),

$$\mathbf{P}(B_n) = \left(\frac{5}{6}\right)^n,$$

for arbitrary n. We denote by B the event of interest to us, that no six appears in an infinite sequence of rolls. If the event B occurs, then all of the events $B_1, B_2, \ldots, B_n, \ldots$ must occur. Hence, on the basis of the first property of probability:

$$\mathbf{P}(B) \leq \mathbf{P}(B_1) = \frac{5}{6},$$

$$\mathbf{P}(B) \leq \mathbf{P}(B_2) = \left(\frac{5}{6}\right)^2,$$

. . . .

$$\mathbf{P}(B) \leq \mathbf{P}(B_n) = \left(\frac{5}{6}\right)^n,$$

. . . .

The numbers $\frac{5}{6}, \left(\frac{5}{6}\right)^2, \ldots, \left(\frac{5}{6}\right)^n, \ldots$ are the terms of an infinite decreasing geometric progression; these terms will eventually be smaller than any predetermined positive number, provided n is sufficiently large.[1] Hence, $\mathbf{P}(B)$ also becomes smaller than any positive number, that is, $\mathbf{P}(B) = 0$. Hence, the probability that no six appears in an infinite sequence of rolls of a die is equal to zero.

The events that we have dealt with up to this time have either been impossible, or, if possible, had a probability greater than zero. Here we encounter for the first time an event whose probability is equal to zero and which, nevertheless, appears to be logically possible. We could not obtain this result if the probability of our event were calculated by the rule set forth at the beginning of this section, namely, as the ratio of the number of favorable events to the total number of all possible events.

The result that we have obtained as our solution to this exercise can be interpreted in the following way. However high we may wish the degree of certainty to be, a number of rolls can be given for which the six must come up at least once with this certainty.[2] This is the precise meaning of the statement that the six comes up with a certainty of 1 in infinitely many rolls.

3. THE FORMULA FOR COMPLETE PROBABILITY

DEFINITION. *A system of events A_1, A_2, A_3, ..., A_n is called complete if at least one of the events must occur (in other words, if the event $A_1 + A_2 + A_3 + \cdots + A_n$ is certain).*

If the events $A_1, A_2, \ldots, A_n$ form a *complete* system and if they are mutually exclusive, then

$$\mathbf{P}(A_1) + \mathbf{P}(A_2) + \cdots + \mathbf{P}(A_n) = 1 \qquad (7)$$

(this follows from formula (5) and Property 4).

---

[1] Proofs of this are found in the section on limits in many calculus textbooks.

[2] In fact, if we say that a six occurs in n rolls, we can err with a probability of $(\frac{5}{6})^n$. For sufficiently great n, the probability of an error can be made arbitrarily small.

Property 7. (The formula for complete probability.) Let a complete system of mutually exclusive events $A_1, A_2, A_3, \ldots, A_n$ be given. Then the probability of an arbitrary event B can be calculated by the formula

$$\mathbf{P}(B) = \mathbf{P}(A_1)\mathbf{P}(B|A_1) + \mathbf{P}(A_2)\mathbf{P}(B|A_2) + \cdots + \mathbf{P}(A_n)\mathbf{P}(B|A_n). \quad (8)$$

To prove this, note that since one of the events $A_1, A_2, \ldots, A_n$ occurs, the occurrence of the event B is equivalent to the appearance of one of the events $BA_1, BA_2, \ldots, BA_n$. Hence,

$$\mathbf{P}(B) = \mathbf{P}((A_1B) + (A_2B) + \cdots + (A_nB)).$$

Since the events $A_1B, A_2B, \ldots, A_nB$ are mutually exclusive (as $A_1, A_2, \ldots, A_n$ are mutually exclusive), we have, by formula (5),

$$\mathbf{P}(B) = \mathbf{P}(A_1B) + \mathbf{P}(A_2B) + \cdots + \mathbf{P}(A_nB).$$

Applying Property 6, we obtain

$$\mathbf{P}(B) = \mathbf{P}(A_1)\mathbf{P}(B|A_1) + \mathbf{P}(A_2)\mathbf{P}(B|A_2) + \cdots + \mathbf{P}(A_n)\mathbf{P}(B|A_n).$$

Let us use this formula to calculate the probability that in the game we described on page 8 a member of the first grade begins. Here, B means that a member of the first grade begins, A_1 that a boy begins, and A_2 that a girl begins. We obtain

$$\mathbf{P}(A_1) = \frac{10}{17}, \ \mathbf{P}(A_2) = \frac{7}{17}, \ \mathbf{P}(B|A_1) = \frac{8}{10}, \ \mathbf{P}(B|A_2) = \frac{3}{7},$$

$$\mathbf{P}(B) = \mathbf{P}(A_1)\mathbf{P}(B|A_1) + \mathbf{P}(A_2)\mathbf{P}(B|A_2)$$

$$= \frac{10}{17}\cdot\frac{8}{10} + \frac{7}{17}\cdot\frac{3}{7} = \frac{8}{17} + \frac{3}{17} = \frac{11}{17}.$$

This result agrees with the previous calculation.

Problem 4. Two players alternately toss a coin, and the one that first tosses heads wins. What is the probability that the game never ends? What is the probability that the first player wins? What is the probability that the other player wins?

Problem 5. A particle at point A (Fig. 4) can, in the next unit of time, remain at A with the probability p_{11}, or move to point B with the probability p_{12}. If it is at B, it can, in the next unit of time,

remain at B with the probability p_{22} or
move to point A with the probability p_{21}.
What is the probability that the particle
is at point A after n units of time, if at
the initial moment it is:

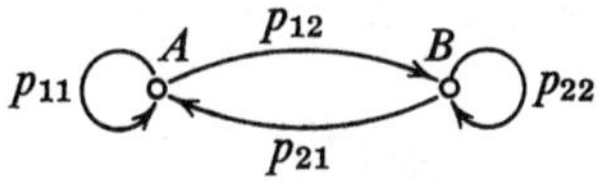

Fig. 4

(*a*) At point A?   (*b*) At point B?

Many different phenomena can be reduced to the diagram in Figure 4. Following the example of the distinguished Russian mathematician A. A. Markov, we consider the first 20,000 letters of the poem *Eugene Onegin* (except ъ and ь). Our particle is at point A if the letter is a vowel and at point B if the letter is a consonant. The succession of vowels and consonants is represented by the motion of the particle on the diagram shown in Figure 4. Obviously, the probability that the letter following a vowel is a consonant is greater than that the letter is once more a vowel. In fact, Markov's calculations show that the probability p_{12} for the appearance of a consonant under the condition that the previous letter was a vowel is approximately equal to 0.872, while the probability p_{11} for the appearance of a vowel under the same condition is equal to about 0.128. It was shown in the same way that $p_{21} \approx 0.663$ and $p_{22} \approx 0.337$.

Similar counts made of the first 100,000 letters of Aksakov's story *The Childhood of Bagrov's Grandson* give a somewhat different result:

$$p_{11} \approx 0.147; \quad p_{21} \approx 0.695;$$
$$p_{12} \approx 0.853; \quad p_{22} \approx 0.305.$$

With certain limitations, a number of meteorological phenomena can also be handled in a similar fashion, for example, the sequence of clear and cloudy days. The probability that a cloudy day will follow a cloudy day is greater than the probability that the following day will be clear; the probability that a clear day will follow a clear day is greater than the probability that the following day will be cloudy. The probability of a change from a clear day to a cloudy one and from a cloudy to a clear one, etc. (a movement of the particle in Problem 5 corresponds to this change), proves to be approximately constant for a definite place and season and can be calculated from observations.

Problem 6. Two identical-looking urns stand in a room. Suppose there are a balls in the left urn and b in the right. Several people come into the room, one after the other, and either transfer a ball from the right urn to the left or from the left urn to the right. It is assumed that the probability that a ball is transferred from the right urn to the left is equal to the probability that a ball is transferred from the left urn to the right, that is, $\frac{1}{2}$. The experiment goes on until one of the urns is empty. What is the probability that the left urn becomes empty? What is the probability that the right urn becomes empty? What is the probability that the experiment does not end?

Problem 7. A caterpillar crawls along the edges of a wire cube (Fig. 5). On reaching a corner, the probability that it will crawl onto any particular edge that leads out from this corner is $\frac{1}{3}$. The points A and B are daubed with glue. The caterpillar starts out from the point 0. What is the probability that it sticks to the point A? What is the probability that it sticks to the point B?

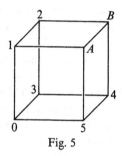

Fig. 5

2. Problems Concerning
a Random Walk on
an Infinite Line

4. GRAPH OF COIN TOSSES

We mark the points 0, ±1, ±2, ±3, ... on a straight line and carry out the following experiment: We place a marker on the point 0 and toss a coin. If heads comes up, we move the marker one place to the left, and if tails comes up, we move the marker one place to the right. Now we toss the coin for a second time, for a third time, etc., and each time move the piece according to the result of the toss. We can assume that the two possible outcomes of a toss have equal probabilities, so that for each toss the probability that the marker is moved to the left is exactly as great as the probability that the marker is moved to the right; that is, the probability is $\frac{1}{2}$.

Clearly, after the first toss, the marker is on the point −1 or 1; after the second toss, on one of the points −2, 0, or 2; after the third, on −3, −1, 1, 3, etc. The diagram shown in Figure 6 gives a graphical picture of the possible positions of the marker at each moment.

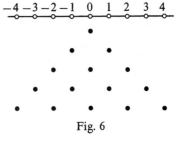

Fig. 6

The points shown in this diagram form a triangle. (This triangle can be continued downwards indefinitely.) The vertex of the triangle lies under the number 0, corresponding to the fact that 0 was the initial position of our marker.

The first row of the triangle consists of two points. The numbers lying over these points, −1 and 1, show where the marker can stand after the first toss. The following row shows where the marker can stand after the second toss, and so on.

5. THE TRIANGLE OF PROBABILITIES

The diagram shown in Figure 6 exhibits all of the possible positions of the marker, but of these some are more, others less probable. We seek to calculate these probabilities. At the beginning, the marker is on the point 0 with probability 1. After the coin has been tossed for the first time, the marker is found with a probability of $\frac{1}{2}$ on each of the points -1 and 1. After the second toss, one can have obtained the following results:

heads–heads,   heads–tails,   tails–heads,   tails–tails.

These four results are all equally probable, and, consequently, each has the probability $\frac{1}{4}$. After the first result, the marker is on point -2; after the second and third, on 0; and after the fourth, on $+2$. Hence, after the first two tosses the marker is found on the point -2 with a probability of $\frac{1}{4}$; on the point 0 with a probability of $\frac{2}{4}$; and on the point $+2$ with a probability of $\frac{1}{4}$. Similarly, the probability of each possible position of the marker after the third, fourth, ... toss can be calculated. If one replaces every point of the diagram by the corresponding probability, one obtains the triangle of numbers shown in Figure 7. This triangle (we shall call it the *triangle of probabilities*) has a noteworthy property: *each of its numbers is equal to half of the sum of the two numbers standing above it.* This property can be easily verified for all of the numbers shown in Figure 7. It is clear, however, that this check still does not prove that this property continues to hold for an arbitrary continuation of the triangle.

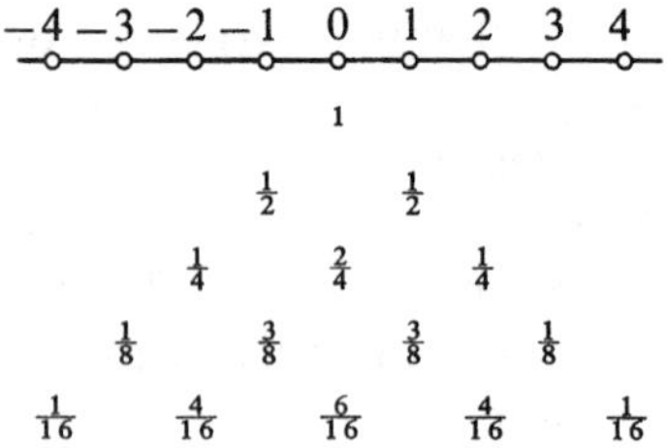

Fig. 7

Problem 8. Prove, with the help of the formula for complete probability (Property 7, page 14), that

$$Z_n{}^k = \frac{1}{2}(Z_{n-1}{}^{k-1} + Z_{n-1}{}^{k+1}), \tag{1}$$

where $Z_n{}^k$ denotes the probability that the marker is at the point k at time n.

With the help of the *rule of the half-sum,* the triangle of probabilities can be easily continued. The first nine rows of the triangle of probabilities are written out in Figure 8 (not counting the zeroth

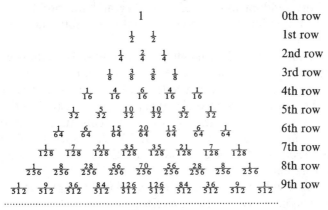

1	0th row
$\frac{1}{2}$ $\frac{1}{2}$	1st row
$\frac{1}{4}$ $\frac{2}{4}$ $\frac{1}{4}$	2nd row
$\frac{1}{8}$ $\frac{3}{8}$ $\frac{3}{8}$ $\frac{1}{8}$	3rd row
$\frac{1}{16}$ $\frac{4}{16}$ $\frac{6}{16}$ $\frac{4}{16}$ $\frac{1}{16}$	4th row
$\frac{1}{32}$ $\frac{5}{32}$ $\frac{10}{32}$ $\frac{10}{32}$ $\frac{5}{32}$ $\frac{1}{32}$	5th row
$\frac{1}{64}$ $\frac{6}{64}$ $\frac{15}{64}$ $\frac{20}{64}$ $\frac{15}{64}$ $\frac{6}{64}$ $\frac{1}{64}$	6th row
$\frac{1}{128}$ $\frac{7}{128}$ $\frac{21}{128}$ $\frac{35}{128}$ $\frac{35}{128}$ $\frac{21}{128}$ $\frac{7}{128}$ $\frac{1}{128}$	7th row
$\frac{1}{256}$ $\frac{8}{256}$ $\frac{28}{256}$ $\frac{56}{256}$ $\frac{70}{256}$ $\frac{56}{256}$ $\frac{28}{256}$ $\frac{8}{256}$ $\frac{1}{256}$	8th row
$\frac{1}{512}$ $\frac{9}{512}$ $\frac{36}{512}$ $\frac{84}{512}$ $\frac{126}{512}$ $\frac{126}{512}$ $\frac{84}{512}$ $\frac{36}{512}$ $\frac{9}{512}$ $\frac{1}{512}$	9th row

Fig. 8

row, the vertex of the triangle). Their direct calculation (by the method used for the first four rows) would not be easy.

Problem 9. Prove that the sum of the elements of each row of the triangle of probabilities is equal to one.

We remark that the *rule of the half-sum* is equivalent to the following *halving-rule.* We consider an arbitrary row of the triangle, halve every number of this row, and place one half below and to the right, the other half below and to the left (see Figure 9, in

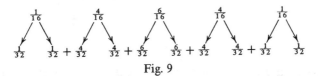

Fig. 9

which this is carried out for the fourth row). If we add the numbers that stand on the same point, we obtain the next row of the triangle.

Let us now imagine that at the initial time there is a unit mass at the point *O,* and that in the course of a second it splits into two equal parts, one half moving to the left and the other half to the right. During the second second these halves again divide into two equal parts, one of which moves to the right, the other to the left, etc.

It is clear that the sum of the masses that are found at the point k after n seconds is equal to the number in our triangle that stands in the kth place of the nth row. This connection between the problem of the random motion of a marker and, the problem of the shifting of a dividing mass is very useful in the solution of a number of problems.[1]

If the first row of the triangle of probabilities is multiplied by 2, the second by 2^2, the third by $2^3, \ldots$, the nth by 2^n, we obtain a triangle that consists only of whole numbers. The reader can easily verify that, in this new triangle, every number is equal to the sum of the two numbers standing above it. This triangle is called *Pascal's triangle;* see E. B. Dynkin and V. A. Uspenskii, *Problems in the Theory of Numbers* (Boston: D. C. Heath and Company, 1963), Chapter 3, section 18.

Let us divide every element of the triangle of probabilities by its left neighbor. Of course, the elements of the left edge of the triangle have no left neighbors. We strike out these elements and what remains is the quotient triangle in Figure 10. The law by

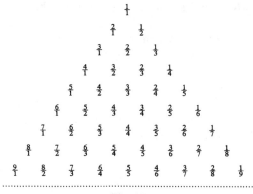

Fig. 10

which this triangle is constructed is easy to discover. In the nth row, the denominators of the fractions run through all the numbers from 1 to n, while the numerators run from n to 1. We leave it to the reader to verify this law with the aid of formula (1).

---

[1] An analogous scheme was considered in one of the problems of the Eighth Moscow Mathematical Olympiad. This exercise dealt with 2^n men who start out from the vertex of the triangle in Figure 6, half of them going down and to the left, and half going down and to the right. At each point, they continue according to the *halving-rule*. It was asked how many people there are at each point of the nth row.

The triangle of probabilities can easily be recovered from the quotient triangle. We begin the nth row with the number $\frac{1}{2^n}$ and then reconstruct all of the elements of this row, one after the other, by multiplying the element already obtained by the corresponding element of the nth row of the quotient triangle. We thus obtain the triangle of probabilities in the form given in Figure 11.

$$1$$

$$\tfrac{1}{2} \qquad \tfrac{1}{2}\cdot\tfrac{1}{1}$$

$$\tfrac{1}{4} \qquad \tfrac{1}{4}\cdot\tfrac{2}{1} \qquad \tfrac{1}{4}\cdot\tfrac{2}{1}\cdot\tfrac{1}{2}$$

$$\tfrac{1}{8} \qquad \tfrac{1}{8}\cdot\tfrac{3}{1} \qquad \tfrac{1}{8}\cdot\tfrac{3}{1}\cdot\tfrac{2}{2} \qquad \tfrac{1}{8}\cdot\tfrac{3}{1}\cdot\tfrac{2}{2}\cdot\tfrac{1}{3}$$

$$\tfrac{1}{16} \qquad \tfrac{1}{16}\cdot\tfrac{4}{1} \qquad \tfrac{1}{16}\cdot\tfrac{4}{1}\cdot\tfrac{3}{2} \qquad \tfrac{1}{16}\cdot\tfrac{4}{1}\cdot\tfrac{3}{2}\cdot\tfrac{2}{3} \qquad \tfrac{1}{16}\cdot\tfrac{4}{1}\cdot\tfrac{3}{2}\cdot\tfrac{2}{3}\cdot\tfrac{1}{4}$$

Fig. 11

From this, it can be seen that the $(k + 1)$st element of the nth row of the triangle of probabilities is equal to

$$\frac{1}{2^n}\cdot\frac{n}{1}\cdot\frac{n-1}{2}\cdot\frac{n-2}{3}\cdot\ \ldots\ \cdot\frac{n-k+1}{k}. \tag{2}$$

6. CENTRAL ELEMENTS OF THE TRIANGLE OF PROBABILITIES

We are particularly interested in the central elements of the triangle of probabilities, that is, the elements that lie on its axis of symmetry. These elements are found only on even rows. The central element $Z_{2k}{}^0$, which lies on the $2k$th row, is the probability that the marker will have returned to the initial position after $2k$ moves. Denoting the probability of this event by w_{2k}, we have

$$w_0 = 1,\ w_2 = \frac{2}{4},\ w_4 = \frac{6}{16},\ w_6 = \frac{20}{64},\ w_8 = \frac{70}{256},\ \ldots,$$

or, by Figure 11,

$$w_0 = 1,\ w_2 = \frac{1}{4}\cdot\frac{2}{1},\ w_4 = \frac{1}{16}\cdot\frac{4}{1}\cdot\frac{3}{2},\ \ldots.$$

If we use the general expression (2) and the fact that the $2k$th row consists of $2k + 1$ numbers, of which k are to the right and k to the left of the middle term, we obtain the formula

$$w_{2k} = \frac{1}{2^{2k}} \cdot \frac{2k}{1} \cdot \frac{2k-1}{2} \cdot \frac{2k-2}{3} \cdot \ldots \cdot \frac{k+1}{k}. \tag{3}$$

The value of w_{2k} can be easily calculated using this formula, provided k is not too large. If k is very large, it is extremely difficult to calculate this fraction. (Try, for example, to calculate $w_{10,000}$!) We can estimate w_{2k} by making use of the following remarkable inequality:

$$\frac{1}{\sqrt{4k}} \leq w_{2k} < \frac{1}{\sqrt{2k}} \quad (k = 1, 2, 3, \ldots). \tag{4}$$

To prove this inequality, we first transform formula (3):

$$w_{2k} = \frac{1}{2^{2k}} \cdot \frac{2k}{1} \cdot \frac{2k-1}{2} \cdot \frac{2k-2}{3} \cdot \ldots \cdot \frac{k+1}{k}$$

$$= \frac{1}{2^{2k}} \cdot \frac{2k}{1} \cdot \frac{2k-1}{2} \cdot \frac{2k-2}{3} \cdot \ldots \cdot \frac{k+1}{k} \cdot \frac{k}{k} \cdot \frac{k-1}{k-1} \cdot \ldots \cdot \frac{1}{1}$$

$$= \frac{1}{2^{2k}} \cdot \frac{1 \cdot 3 \cdot 5 \cdot \ldots \cdot (2k-1) \cdot 2 \cdot 4 \cdot 6 \cdot \ldots \cdot 2k}{1 \cdot 2 \cdot 3 \cdot \ldots \cdot k \cdot 1 \cdot 2 \cdot 3 \cdot \ldots \cdot k}$$

$$= \frac{1}{2^{2k}} \cdot \frac{1}{1} \cdot \frac{3}{2} \cdot \frac{5}{3} \cdot \ldots \cdot \frac{2k-1}{k} \cdot \frac{2}{1} \cdot \frac{4}{2} \cdot \frac{6}{3} \cdot \ldots \cdot \frac{2k}{k}$$

$$= \frac{1}{2^k} \cdot \frac{1}{1} \cdot \frac{3}{2} \cdot \frac{5}{3} \cdot \ldots \cdot \frac{2k-1}{k} = \frac{1}{2} \cdot \frac{3}{4} \cdot \frac{5}{6} \cdot \ldots \cdot \frac{2k-1}{2k}.$$

We now write the three products

$$\frac{1}{2} \cdot \frac{1}{2} \cdot \frac{2}{3} \cdot \frac{3}{4} \cdot \frac{4}{5} \cdot \frac{5}{6} \cdot \frac{6}{7} \cdot \frac{7}{8} \cdot \frac{8}{9} \cdot \frac{9}{10} \cdot \ldots \cdot \frac{2k-2}{2k-1} \cdot \frac{2k-1}{2k}, \tag{5}$$

$$\frac{1}{2} \cdot \frac{1}{2} \cdot \frac{3}{4} \cdot \frac{3}{4} \cdot \frac{5}{6} \cdot \frac{5}{6} \cdot \frac{7}{8} \cdot \frac{7}{8} \cdot \frac{9}{10} \cdot \frac{9}{10} \cdot \ldots \cdot \frac{2k-1}{2k} \cdot \frac{2k-1}{2k}, \tag{6}$$

$$\frac{1}{2} \cdot \frac{2}{3} \cdot \frac{3}{4} \cdot \frac{4}{5} \cdot \frac{5}{6} \cdot \frac{6}{7} \cdot \frac{7}{8} \cdot \frac{8}{9} \cdot \frac{9}{10} \cdot \frac{10}{11} \cdot \ldots \cdot \frac{2k-1}{2k} \cdot \frac{2k}{2k+1} \tag{7}$$

under one another. It is easily seen that of the three numbers in any column, the second is at least equal to the first, and the third is at least equal to the second. Hence, (5) has the smallest and (7) the greatest value. The middle product is equal to

$$\left(\frac{1}{2}\right)^2 \cdot \left(\frac{3}{4}\right)^2 \cdot \left(\frac{5}{6}\right)^2 \cdot \ldots \cdot \left(\frac{2k-1}{2k}\right)^2 = w_{2k}^2,$$

the upper, after simplification, is equal to $\frac{1}{4k}$, and the lower is equal to $\frac{1}{2k+1}$. Hence,

$$\frac{1}{4k} \leq w_{2k}^2 \leq \frac{1}{2k+1}.$$

Then, certainly,

$$\frac{1}{4k} \leq w_{2k}^2 < \frac{1}{2k},$$

or, if we take the square root,

$$\frac{1}{\sqrt{4k}} \leq w_{2k} < \frac{1}{\sqrt{2k}},$$

which was to be proved.

By the same method, still more exact approximations for w_{2k} can be found.

Problem 10. Prove that, for all $k \geq 2$, we have

$$\sqrt{3 \cdot \left(\frac{1}{2}\right)^2 \cdot \frac{3}{4} \cdot \frac{1}{2k}} \leq w_{2k} < \sqrt{3 \cdot \left(\frac{1}{2}\right)^2 \cdot \frac{1}{2k}}; \qquad (8)$$

for all $k \geq 3$, we have

$$\sqrt{5 \cdot \left(\frac{1}{2} \cdot \frac{3}{4}\right)^2 \cdot \frac{5}{6} \cdot \frac{1}{2k}} \leq w_{2k} < \sqrt{5 \cdot \left(\frac{1}{2} \cdot \frac{3}{4}\right)^2 \cdot \frac{1}{2k}}; \qquad (9)$$

and, in general, for all $k \geq a$

$$\sqrt{(2a-1) \cdot \left(\frac{1}{2} \cdot \frac{3}{4} \cdot \ldots \cdot \frac{2a-3}{2a-2}\right)^2 \cdot \frac{2a-1}{2a} \cdot \frac{1}{2k}}$$

$$\leq w_{2k} < \sqrt{(2a-1) \cdot \left(\frac{1}{2} \cdot \frac{3}{4} \cdot \ldots \cdot \frac{2a-3}{2a-2}\right)^2 \cdot \frac{1}{2k}}. \qquad (10)$$

Hint. In each of the products (5) and (7), change the initial terms in such a way that they coincide with the initial terms of the product (6).

Problem 10 yields a sequence of increasingly accurate approximations for w_{2k}; the ratio of the lower to the upper bound serves as a measure of the accuracy. For approximation (4), this is equal to $\sqrt{\frac{1}{2}}$; for approximation (8), it is equal to $\sqrt{\frac{3}{4}}$; for approximation (9), it is equal to $\sqrt{\frac{5}{6}}$; and for approximation (10), it is equal to $\sqrt{1 - \frac{1}{2a}}$, and, thus, approaches 1 as a limit. If the products in the inequalities (8) and (9) are calculated out, we obtain

$$\frac{1}{\sqrt{3.56k}} \leq w_{2k} < \frac{1}{\sqrt{2.66k}} \quad \text{(for } k \geq 2\text{)}, \tag{11}$$

$$\frac{1}{\sqrt{3.42k}} \leq w_{2k} < \frac{1}{\sqrt{2.84k}} \quad \text{(for } k \geq 3\text{)}. \tag{12}$$

On substituting the values 4, 5, 60, 150 for a in the general inequality (10), we obtain

$$\frac{1}{\sqrt{3.35k}} \leq w_{2k} < \frac{1}{\sqrt{2.92k}} \quad \text{(for } k \geq 4\text{)}, \tag{13}$$

$$\frac{1}{\sqrt{3.31k}} \leq w_{2k} < \frac{1}{\sqrt{2.98k}} \quad \text{(for } k \geq 5\text{)}, \tag{14}$$

$$\frac{1}{\sqrt{3.18k}} \leq w_{2k} < \frac{1}{\sqrt{3.10k}} \quad \text{(for } k \geq 60\text{)}, \tag{15}$$

$$\frac{1}{\sqrt{3.15k}} \leq w_{2k} < \frac{1}{\sqrt{3.14k}} \quad \text{(for } k \geq 150\text{)}. \tag{16}$$

The coefficients of k constantly decrease on the left side of the inequality, while they constantly increase on the right side. One can prove rigorously (we shall not do so here) that both of these sequences converge to the same limit, and that this limit is the well-known π (the ratio of the circumference of a circle to its diameter). Hence, w_{2k} can be calculated for large values of $2k$ by the following approximation formula:

$$w_{2k} \approx \frac{1}{\sqrt{\pi k}}. \tag{17}$$

One can prove (and the reader can verify this for himself) that for $k = 25$ this formula yields an approximation that is correct to two significant figures; the larger the value of k, the more accurate the approximation.

Problem 11. Calculate $w_{10,000}$ to two significant figures.

In the triangle of probabilities, the numbers increase as we move from the edges to the middle. Hence, by using the inequality (4) it is not difficult to find the upper bound of the numbers in the nth row.

Problem 12. Prove that all elements of the nth row of the triangle of probabilities are less than or equal to $\dfrac{1}{\sqrt{n}}$.

7. ESTIMATION OF ARBITRARY ELEMENTS OF THE TRIANGLE

Since we have found an approximation formula for the terms lying in the middle of the triangle of probabilities, it is only natural to seek an equally convenient formula for the other terms of the triangle.

Consider the $2k$th row. We denote the middle term of this row by v_0 (above, we called this term w_{2k}) and number all of the terms to its right in this row

$$v_0, v_1, v_2, \ldots, v_{k-2}, v_{k-1}, v_k.$$

Now, the elements on the right side of the $2k$th row of the quotient triangle are

$$\frac{k}{k+1}, \frac{k-1}{k+2}, \cdots, \frac{2}{2k-1}, \frac{1}{2k}.$$

By the definition of the quotient triangle, we have

$$\frac{v_1}{v_0} = \frac{k}{k+1}; \; \frac{v_2}{v_1} = \frac{k-1}{k+2}; \; \cdots; \; \frac{v_{k-1}}{v_{k-2}} = \frac{2}{2k-1}; \; \frac{v_k}{v_{k-1}} = \frac{1}{2k}.$$

If we choose a number s between 0 and k, and estimate the ratio $\frac{v_s}{v_0}$, we obtain

$$\frac{v_s}{v_0} = \frac{v_1}{v_0} \cdot \frac{v_2}{v_1} \cdot \ldots \cdot \frac{v_{s-1}}{v_{s-2}} \cdot \frac{v_s}{v_{s-1}}$$

$$= \frac{k}{k+1} \cdot \frac{k-1}{k+2} \cdot \ldots \cdot \frac{k-s+2}{k+s-1} \cdot \frac{k-s+1}{k+s}.$$

If we reverse the order of the denominators appearing in these factors, we get

$$\frac{v_s}{v_0} = \frac{k}{k+s} \cdot \frac{k-1}{k+s-1} \cdot \ldots \cdot \frac{k-s+2}{k+2} \cdot \frac{k-s+1}{k+1}$$

$$= \left(1 - \frac{s}{k+s}\right)\left(1 - \frac{s}{k+s-1}\right) \cdots \left(1 - \frac{s}{k+2}\right)\left(1 - \frac{s}{k+1}\right).$$

It is easy to see that the first factor is greater than each of those following it, while the last factor is less than each of those preceding it. It follows that

$$\left(1 - \frac{s}{k+1}\right)^s \leq \frac{v_s}{v_0} \leq \left(1 - \frac{s}{k+s}\right)^s, \tag{18}$$

or,

$$\left(\frac{k+1-s}{k+1}\right)^s \leq \frac{v_s}{v_0} \leq \left(\frac{k}{k+s}\right)^s. \tag{19}$$

Problem 13. Between what limits does the probability $Z_{120}{}^{20}$ lie?

One can find analogous estimates for the elements in the odd-numbered rows; however, we shall not concern ourselves with this.

8. THE LAW OF THE SQUARE ROOT OF n

Assume that our experiment of tossing the coin and moving the marker is continued for sufficiently long, say for 1,000 moves. How far is the marker then from the starting point? In any case, not farther than 1,000 steps to the right or left. And this is the only thing that we can state with absolute certainty. If we seek to assert that the marker moves less than 1,000 steps, that is, not more than 998 steps, then an error is not impossible, for the same side of the coin can come up every time in 1,000 tosses (heads or tails), and the marker will then be found at $-1,000$ or at $+1,000$. The probability that this happens is, however, so small $\left(\text{it is equal to } 2 \cdot \dfrac{1}{2^{1,000}}\right)$ that we can regard such an outcome as practically impossible. Likewise, the probability that the marker passes the nine hundred ninetieth mark is small. For this to happen, the marker must reach one of the points

$$-1,000, \ -998, \ -996, \ -994, \ -992,$$
$$+992, \ +994, \ +996, \ +998, \ +1,000.$$

To calculate the probability of this, we must consider the thousandth row of the triangle of probabilities and form the sum of the five outer left and the five outer right terms. Because of the symmetry of the triangle, this sum is equal to twice the sum of the five outer left or right terms:

$$2\left(\frac{1}{2^{1,000}} + \frac{1}{2^{1,000}} \cdot \frac{1,000}{1} + \frac{1}{2^{1,000}} \cdot \frac{1,000}{1} \cdot \frac{999}{2}\right.$$
$$+ \frac{1}{2^{1,000}} \cdot \frac{1,000}{1} \cdot \frac{999}{2} \cdot \frac{998}{3}$$
$$\left.+ \frac{1}{2^{1,000}} \cdot \frac{1,000}{1} \cdot \frac{999}{2} \cdot \frac{998}{3} \cdot \frac{997}{4}\right) \approx 0.\underbrace{00 \ldots 0}_{290 \text{ zeros}} \ldots 8.$$

Hence, we can regard an error here as practically impossible. This is not surprising, for when it is assumed that after one thousand moves the marker has not passed the 990th mark, of the 1,001 possible positions we neglect the ten most improbable ones. The statement that the marker remains under the hundredth mark (that is, that it does not pass the ninety-eighth mark) is much more risky. Here, we neglect the greater part of the possible positions, namely, 902 out of 1,001. How justified are we in doing this? How small is the probability of an error in this statement or, on the other hand, how close to one is it? This question cannot be answered without calculation. To carry it through, we must find the sum of the 451 left outer and the 451 right outer terms in the thousandth row of the triangle of probabilities. While no knowledge is necessary for this calculation except that of the four fundamental arithmetic operations, few of our readers could carry it through to the end. It would take too much time and energy.[1] We shall there-fore seek to estimate the probability, and forego an exact calcula-tion. Our considerations will be altogether general. To simplify the calculations, however, we shall assume from now on that the num-ber n of steps that the marker makes is even.

Let us consider the $2k$th row of the triangle of probabilities and number its terms from the middle term to the right edge

$$v_0, v_1, v_2, \ldots, v_k.$$

To estimate the sum

$$S = v_r + v_{r+1} + \cdots + v_k,$$

we use inequality (19) of section 7.

By this inequality, we have

$$\frac{v_r}{v_0} \leq \left(\frac{k}{k+r}\right)^r,$$

$$\frac{v_{r+1}}{v_0} \leq \left(\frac{k}{k+r+1}\right)^{r+1},$$

$$\frac{v_{r+2}}{v_0} \leq \left(\frac{k}{k+r+2}\right)^{r+2},$$

$$\cdots \cdots \cdots \cdots \cdots$$

$$\frac{v_k}{v_0} \leq \left(\frac{k}{k+k}\right)^k.$$

For brevity, we denote $\dfrac{k}{k+r}$ by g. One sees immediately that no fraction enclosed in parentheses on the right-hand sides of our inequalities exceeds g. Hence,

$$\frac{v_r}{v_0} \leq g^r, \quad \frac{v_{r+1}}{v_0} \leq g^{r+1}, \quad \frac{v_{r+2}}{v_0} \leq g^{r+2}, \quad \cdots, \quad \frac{v_k}{v_0} \leq g^k.$$

---

[1] Naturally, we can calculate this sum in the following way: We subtract from one the sum of the 99 middle terms. The number of summands in this sum is significantly smaller (99 instead of 902), but these summands themselves are considerably more difficult to calculate.

Adding these inequalities, we obtain

$$\frac{S}{v_0} \le g^r + g^{r+1} + g^{r+2} + \cdots + g^k.$$

The right-hand side forms a geometric series, whose sum is $\dfrac{g^r - g^{k+1}}{1 - g}$; thus,

$$\frac{S}{v_0} \le \frac{g^r - g^{k+1}}{1 - g}.$$

But, $g = \dfrac{k}{k + r} < 1$, so that $1 - g > 0$, and the inequality is strengthened when we omit the negative number $\dfrac{-g^{k+1}}{1 - g}$ from the right-hand side. Hence,

$$\frac{S}{v_0} < \frac{g^r}{1 - g}.$$

If we multiply this inequality by v_0 and substitute the original value for g, we finally obtain

$$S < v_0 \frac{k + r}{r} \left(\frac{k}{k + r} \right)^r. \tag{20}$$

Now we are in a position to estimate the probability P that after $n = 2k$ moves, the marker is not less than $m = 2r$ steps from the starting point. This probability is equal to twice the probability that after $n = 2k$ moves, the marker is not less than $m = 2r$ steps to the right of the starting point. The probability of this last event is precisely the sum S, which we have already estimated. In formula (20) we replace v_0 (for the middle term of the $2k$th row) by the more convenient symbol w_{2k}, used previously; this reminds us that the middle term depends on the number of the row. We multiply both sides of inequality (20) by 2, replace k by $\dfrac{n}{2}$ and r by $\dfrac{m}{2}$, and thus obtain

$$P \le 2w_n \cdot \frac{n + m}{m} \left(\frac{n}{n + m} \right)^{\frac{m}{2}}. \tag{21}$$

The middle term

$$w_n = w_{2k}$$

can be estimated by either formula (4) or (11)–(16) of section 6. By these formulas, we have

$$w_{2k} < \frac{1}{\sqrt{Bk}}, \tag{22}$$

where B is a number which can be chosen as near to π as desired, provided k is large enough. (In (4) $B = 2$; in (11) $B = 2.66$, etc.)

We replace k by $\frac{n}{2}$ in the inequality (22). Comparing this inequality with (21), we obtain

$$P \leq 2 \sqrt{\frac{2}{Bn}} \cdot \frac{n+m}{m} \cdot \left(\frac{n}{n+m}\right)^{\frac{m}{2}}, \tag{23}$$

and can use this formula to calculate the numerical example discussed at the beginning of this section. Suppose the marker makes 1,000 moves. What is the probability that it is not less than 100 steps from the starting point? We substitute $n = 1,000$ and $m = 100$ in formula (23):

$$P \leq 2 \sqrt{\frac{2}{1,000B}} \cdot 11 \cdot \left(\frac{10}{11}\right)^{50}.$$

Since $k = \frac{n}{2} > 150$, we can set $B = 3.14$ according to formula (16) of section 6. We then obtain

$$P < 0.0048.$$

The event that the marker is at least 100 steps from the starting point after 1,000 moves is thus highly improbable. If a not too high degree of certainty is demanded and, say, 0.005 is permitted as the degree of uncertainty, one can say that it is practically impossible that the marker has moved more than 98 steps after 1,000 moves; that is, it is practically certain that the marker has moved less than 100 steps.

For the further study of the motion of the marker, it is convenient to replace the approximation (23) by another that is not so precise but is simpler and more convenient to calculate. We require a preliminary inequality, the proof of which is left to the reader.

Problem 14. Prove that for arbitrary positive p and integral positive r,

$$(1 + p)^r \geq 1 + rp. \tag{24}$$

Let m and n be positive even numbers. We substitute $r = \dfrac{m}{2}$ and $p = \dfrac{m}{n}$ in the inequality (24) and obtain

$$\left(1 + \frac{m}{n}\right)^{\frac{m}{2}} \geq 1 + \frac{m^2}{2n} > \frac{m^2}{2n}.$$

Hence,

$$\left(\frac{n}{n + m}\right)^{\frac{m}{2}} = \frac{1}{\left(1 + \dfrac{m}{n}\right)^{\frac{m}{2}}} < \frac{2n}{m^2}. \tag{25}$$

From inequalities (21) and (25) it follows that

$$P < 2w_n(n + m)\frac{2n}{m^3}. \tag{26}$$

According to inequality (4) in section 6, $w_n < \dfrac{1}{\sqrt{n}}$; furthermore, $m \leq n$. Hence, the estimate

$$P < \frac{2}{\sqrt{n}} \cdot 2n \cdot \frac{2n}{m^3} = \left(\frac{2\sqrt{n}}{m}\right)^3$$

follows from inequality (26).

Thus, if it is asserted that the marker is less than m steps distant from the starting point after n moves, the probability of error is less than $\left(\dfrac{2\sqrt{n}}{m}\right)^3$.

We choose an arbitrary positive number t and estimate the probability of an error in the following statement:

(A) After n moves the marker is less than $t\sqrt{n}$ steps away from the starting point.

We denote by m the smallest even number that satisfies the condition

$$m \geq t\sqrt{n}.$$

Since the distance of the marker from the starting point after an even number of moves is an even number, the statement (A) is equivalent to the following statement:

(B) After n moves the marker is less than m steps away from the starting point.

Consequently, the probability of an error in the statement (A) is equal to the probability of an error in the statement (B). This probability is smaller than

$$\left(\frac{2\sqrt{n}}{m}\right)^3 \le \left(\frac{2\sqrt{n}}{t\sqrt{n}}\right)^3 = \left(\frac{2}{t}\right)^3.$$

Hence, we have proved the following important law:

LAW (The Law of the Square Root of n). *With probability of error less than* $\left(\frac{2}{t}\right)^3$, *one can assert that after n moves, the distance of the marker from the starting point is less than* $t\sqrt{n}$ *(that is, the marker is situated between* $-t\sqrt{n}$ *and* $+t\sqrt{n}$*).*

We choose a certain degree of uncertainty, for example 0.005, and determine t so that

$$\left(\frac{2}{t}\right)^3 = 0.005.$$

As a solution of this equation for t, we find[1]

$$t = \frac{2}{\sqrt[3]{0.005}} \approx 12.$$

It follows from the *law of the square root of n* that, for every value n, the statement that *the marker has moved less than* $12\sqrt{n}$ *steps from the starting point in n moves* is practically certain.

We compile the following table:

n	$n' = 12\sqrt{n}$	$\dfrac{n'}{n} = \dfrac{12}{\sqrt{n}}$
2,500	600	0.24
10,000	1,200	0.12
250,000	6,000	0.024
1,000,000	12,000	0.012

[1] Here the approximation is greater than the actual value.

The second column of the table gives almost certain bounds on the distance of the marker from the starting point for various values of n. The ratio $\dfrac{n'}{n} = \dfrac{12}{\sqrt{n}}$ approaches zero as n increases without bound.

Let us now assume that the marker is m steps distant from the starting point at the end of n moves. We call the ratio $\dfrac{m}{n}$ the *reduced velocity* of the marker. If a particle starts out from the point 0 with this velocity and does not change direction, then at time n, its displacement amounts to m steps.

For example: If the marker is at the point -20 after 100 steps, its *reduced velocity* amounts to $\frac{1}{5}$. The reduced velocity varies between 1 (when the marker always moves in one direction) and 0 (when it returns to the starting point, that is, when it makes exactly as many moves to the left as to the right). From the *law of the square root of n* one easily deduces:

THEOREM. *When the motion of the marker is continued sufficiently long, it is practically certain that the reduced velocity is close to zero.*

In fact, it is practically certain that the displacement of the marker is less than $12\sqrt{n}$, and, hence, that its reduced velocity is less than $\dfrac{12\sqrt{n}}{n} = \dfrac{12}{\sqrt{n}}$. If n is sufficiently large, this bound will be arbitrarily close to zero.

Until now, we have taken 0.005 as the permissible probability of error. However, we can repeat our considerations without significant alteration for an arbitrary value ε of this tolerance. As a result, we come to the following conclusion:

One can say with probability of error less than ε that:

(*a*) The displacement of the marker after n moves is less than

$$\dfrac{2}{\sqrt[3]{\varepsilon}}\sqrt{n}\,;$$

(*b*) The absolute value of its reduced velocity after n moves is less than $\dfrac{2}{\sqrt[3]{\varepsilon}}\Big/\sqrt{n}$.

Problem 15. Let the permissible probability of error be 0.05. Give a practically certain bound for the displacement and for the reduced velocity of the marker after 1,000 moves.

Problem 16. Let α be an arbitrary positive number. Prove that it can be stated with the probability of error less than

$$\left(\frac{2}{\alpha\sqrt{n}}\right)^3$$

that the reduced velocity of the marker after n moves is less than α.

Problem 17. Determine the number of moves that are sufficient for the reduced velocity of the marker to be smaller than 0.01, with the probability of error not greater than 0.001.

9. THE LAW OF LARGE NUMBERS

We now recall that the marker moves in accordance with the outcome of the toss of a coin. If on n tosses of a coin there are l tails and $n - l$ heads, the marker moves l steps to the right and $n - l$ steps to the left, finally reaching the point

$$l - (n - l) = 2l - n.$$

The reduced velocity of the marker after n moves is given by the absolute value of

$$\frac{2l - n}{n} = 2\frac{l}{n} - 1. \tag{27}$$

The fraction $\frac{l}{n}$ characterizes the *relative frequency* with which tails comes up.

Let a permissible probability of error be given. We know that for large values of n it can be asserted with practical certainty that the reduced velocity is close to zero. It is clear from equality (27) that for a small reduced velocity $2\frac{l}{n}$ is approximately equal to 1, and the relative frequency $\frac{l}{n}$ is consequently near $\frac{1}{2}$. In other words:

If a coin is tossed very often, it is practically certain that the frequency with which heads comes up is approximately equal to $\frac{1}{2}$.

Roughly speaking, it is practically certain that heads comes up exactly as often as tails. A more exact formulation would be:

Choose an arbitrarily permissible probability of error ε and an arbitrarily small number α. If the number of tosses of the coin exceeds

$$N = \frac{1}{\alpha^2 \sqrt[3]{\varepsilon^2}},$$

it can be asserted with a probability of error less than ε that the frequency with which tails comes up differs from ½ by less than α.

The proof for this exact formulation is easily obtained from part (*b*) on page 33. For $n > \dfrac{1}{\alpha^2 \sqrt[3]{\varepsilon^2}}$ we have

$$\frac{1}{\sqrt{n}} < \alpha \sqrt[3]{\varepsilon},$$

and

$$\frac{\dfrac{2}{\sqrt[3]{\varepsilon}}}{\sqrt{n}} < 2\alpha.$$

Hence, the absolute value of the reduced velocity of the marker is less than 2α, with the probability of error less than ε. In this case, the reduced velocity is equal to the absolute value of $\dfrac{2l - n}{n} = 2\dfrac{l}{n} - 1$. Hence, it can be stated with the probability of error less than ε that $2\dfrac{l}{n}$ does not differ from 1 by more than 2α, or in other words, that $\dfrac{l}{n}$ differs from ½ by less than α.

Problem 18. How often must a coin be tossed so that it can be asserted with the probability of error less than 0.01 that the frequency with which tails comes up lies between 0.4 and 0.6?

Suppose now that a die is tossed instead of a coin. How often does the six come up? If the same calculations and arguments are carried through for this new case as for the example of the coin, we obtain the following result: for a great number of tosses, the frequency with which a six comes up lies near ⅙ with practical certainty.

We consider yet another experiment. An urn contains *a* balls, of which *b* are white and the rest black. A ball is drawn *n* times from this urn, and is returned to the urn each time. How often will a white ball be drawn? One can prove that it is practically certain that for sufficiently many trials, the frequency with which a white ball is drawn lies near $\frac{b}{a}$.

We now formulate a general result including all of the above formulations as special cases.

Suppose that an experiment is carried out in which an event A can either occur or not occur (a toss of tails, the roll of a six, drawing a white ball out of the urn, etc.), *and let the probability of the occurrence of the event A be p.* (In our examples p was equal to $\frac{1}{2}$, $\frac{1}{6}$, and $\frac{b}{a}$, respectively.) *Suppose this experiment is repeated many times, the result of each trial not influencing the results of the succeeding ones. Then, for a large number of trials, it is practically certain that the frequency of the event A will be approximately equal to the probability p of this event.*

This general formulation can be made more precise, exactly as in the formulation for the case of the coin.

This result is essentially nothing but a restatement of a well-known theorem of Bernoulli,[1] which sets forth the simplest form of one of the fundamental laws of probability theory, *the law of large numbers.* Here we cannot go into the generalizations of Bernoulli's theorem. We only remark that the most important is due to the Russian mathematician P. L. Chebyshev.

The reader will appreciate the great significance of the law of large numbers. By the statement that the frequency of occurrence of an event A approaches the probability of A with practical certainty for a large series of trials, the law of large numbers makes possible the experimental determination of this probability. In many cases, the experimental method for the determination of a probability is the only possible one. Furthermore, the knowledge of the connection between probability and frequency enables one to draw practical conclusions about the frequency of appearance of an event in a long series of experiments from the theoretically calculated probability of this event. The connection between probability and frequency is fundamental in many applications of probability theory to physics, technology, etc.

---

[1] Jacob Bernoulli (1654–1705), famous Swiss mathematician.

3. Random Walks with Finitely Many States

In the preceding chapter, we considered the simplest example of a random walk, a random walk on a line. The problems posed there were of the following sort: At a given moment, where could the marker be, what is the probability that it is at a given point, and how far from the starting point is it? In this chapter, we consider more complicated schemes for random walks, including, for example, the random stroll through the city and the children's game mentioned in the Introduction. The problems related to these schemes differ somewhat from those investigated in Chapter 2. If an arbitrary point is chosen, we ask whether a particle can ever reach it, and if so, when? The first question can be answered with the aid of a general theorem (p. 46). An exact answer for the second, dealing with the question of the number of moves necessary to reach a given point with a given probability, can be found only in the simplest cases (see Problem 20 to follow). For the general case, we can give only an approximation for the necessary number of moves.

10. RANDOM WALKS ON A FINITE LINE

Let us make a slight change in the scheme for a random walk on a straight line. We place reflecting barriers in the path of the moving particle at the points m_1 and m_2 (see Fig. 12). These barriers

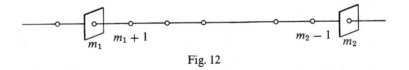

Fig. 12

cause particles that reach m_1 to move to $m_1 + 1$ on the next move, and those that reach m_2 to move to $m_2 - 1$ on the next move. The motion of the particle will thus take place between the points m_1 and m_2.

We could also restrict the motion of the particle by placing absorbing rather than reflecting barriers at the points m_1 and m_2. In this case, the particle upon reaching either the point m_1 or the point m_2 would remain there permanently. (We have already considered this diagram in the solution of Problem 6.)

Finally, we could place a reflecting barrier at one of the two points and an absorbing barrier at the other.

Problem 19. Show that the probability that after n moves the particle has reached the point m at least once does not depend on whether there is a barrier at m, and if there is, whether it is a reflecting or an absorbing barrier.

We place a reflecting barrier at point 0. A particle makes n moves starting from the point 1. What is the probability that it touches the point 3 at least once? To calculate this probability we can, according to Problem 19, place an absorbing barrier at point 3 (Fig. 13). But then, the event that after n moves the particle touches the point 3 at least once is the same as the event that the particle is at

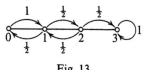

Fig. 13

point 3 after n moves. (For, when it reaches this point it remains there.) We wish to calculate the probability d_n that the particle is at the point 3 after n moves. By a_n, b_n, c_n, we denote the probability of the events that the particle is at point 0, point 1, point 2, respectively.

To be at point 0 after n moves (the probability for this is a_n), the particle must be at point 1 after $n - 1$ moves (the probability for this is b_{n-1}), and then go from 1 to 0 (this occurs with probability $\frac{1}{2}$). By Property 6, we have

$$a_n = b_{n-1} \cdot \frac{1}{2}. \tag{1}$$

Similarly, with the aid of Properties 6 and 7, we obtain the relations

$$b_n = a_{n-1} \cdot 1 + c_{n-1} \cdot \tfrac{1}{2}, \tag{2}$$

$$c_n = b_{n-1} \cdot \tfrac{1}{2}, \tag{3}$$

$$d_n = d_{n-1} \cdot 1 + c_{n-1} \cdot \tfrac{1}{2}. \tag{4}$$

Problem 20. Using relations (1)–(4), show that

$$d_{2k} = d_{2k+1} = 1 - \left(\frac{3}{4}\right)^k. \tag{5}$$

Equality (5) can be interpreted as follows: the statement that after $2k$ moves the particle has touched the point 3 at least once, is false with probability $(\frac{3}{4})^k$. How many moves must be made for this probability to be less than 0.01?

We obtain $k_0 = 16.006$ as a solution of the equation $(\frac{3}{4})^{k_0} = 0.01$. Thus, we conclude that, for

$$k > 16.006, \tag{6}$$

the inequality

$$\left(\frac{3}{4}\right)^k < 0.01 \tag{7}$$

is satisfied. Hence, for every $k \geq 17$, that is, for every number of moves greater than or equal to $2 \cdot 17 = 34$, the particle reaches the point 3 with a probability greater than 0.99. (Compare with the calculations on page 45.)

11. RANDOM WALKS THROUGH A CITY

We return to a consideration of the random walk through a city, which we have described in the Introduction (Fig. 1). We ask whether and when the friends reach the intersection E. We shall prove: If the time of their stroll is without limit, they reach the intersection E, exactly as with all other intersections, with probability 1. Furthermore, we shall estimate the probability that E is reached in a given number of moves.

We state our argument in general form, not assuming that the city must necessarily be of the form shown in Figure 1. Suppose that a traveler goes through the city. If he reaches an intersection from which k streets go out, the probability that he chooses to continue along the first street is p_1, that he chooses the second, p_2, ..., and that he chooses the kth street, p_k (the case that he chooses with a certain probability the street by which he arrived is included in this enumeration). We assume that the numbers $p_1, p_2, \ldots, p_k$ are different from zero, and that for a given intersection the probabilities remain constant. This means that the traveler always chooses his further path from a certain intersection with the same probabilities, independent of the direction from which he entered the intersection and the number of times he has traversed it. (In the example given on page 1, we have $k = 4$ and $p_1 = p_2 = p_3 = p_4 = \frac{1}{4}$ for every intersection.) If the traveler reaches the edge of the city, he turns around and comes back. We shall assume that he continues in this way indefinitely. We claim that, no matter how the traveler wanders, for each intersection, the probability that he comes to this intersection is 1 (regardless of the location of the intersection).

Proof. Part 1. Let E_0 be *any* intersection; we shall show that the probability that the traveler comes to E_0 is 1. Let the remaining intersections be denoted by $E_1, E_2, \ldots, E_v$.

Suppose the positive integer N and the number $\alpha > 0$ are such that the probability of his arriving at E_0 at least once after completing N moves,[1] regardless of his path, is greater than or equal to α.

We split the totality of moves the traveler makes into groups of N moves each. Let the first subsequence consist of the first to the Nth move, the second, the $(N + 1)$st to the $2N$th, etc. We denote by D_k the event that after making all the moves contained in the first k groups, the traveler has come to E_0 at least once. We denote by $\bar{D}_k$ the contrary event (that the traveler has not reached the point E_0 at any time while making the moves contained in the first k groups).

The aim of the first part of the proof is to establish the inequality

$$\mathbf{P}(\bar{D}_k) \leq \mathbf{P}(\bar{D}_{k-1})(1 - \alpha).$$

To do this, we consider the event $F_{k-1}^{(s)}$: "After making all moves in the first $k - 1$ groups, the traveler has not reached E_0 and is at E_s after $(k - 1)N$ moves."

The events $D_{k-1}, F_{k-1}^{(1)}, F_{k-1}^{(2)}, \ldots, F_{k-1}^{(v)}$ are pairwise mutually exclusive; they also form a complete system. By Property 7,

$$\mathbf{P}(D_k) = \mathbf{P}(D_{k-1})\mathbf{P}(D_k|D_{k-1}) + \mathbf{P}(F_{k-1}^{(1)})\mathbf{P}(D_k|F_{k-1}^{(1)}) + \cdots$$

$$+ \mathbf{P}(F_{k-1}^{(v)})\mathbf{P}(D_k|F_{k-1}^{(v)}). \qquad (8)$$

Obviously,

$$\mathbf{P}(D_k|D_{k-1}) = 1. \qquad (9)$$

---

[1] For brevity, we say that the traveler has made a "move" when he goes from one intersection to a neighboring one.

Furthermore, $\mathbf{P}(D_k|F_{k-1}^{(1)})$ is the probability that the traveler reaches the point E_0 with a move in the kth group, upon the condition that he was at E_1 at the beginning of this group. In other words: $\mathbf{P}(D_k|F_{k-1}^{(1)})$ is the probability that E_0 is reached at least once in N moves, if the path begins at E_1. By the choice of α, this probability is greater than or equal to α, that is,

$$\mathbf{P}(D_k|F_{k-1}^{(1)}) \geq \alpha. \tag{10}$$

Likewise,

$$\mathbf{P}(D_k|F_{k-1}^{(2)}) \geq \alpha, \quad \ldots, \quad \mathbf{P}(D_k|F_{k-1}^{(v)}) \geq \alpha. \tag{11}$$

Hence, it follows from Formulas (8)–(11):

$$\mathbf{P}(D_k) \geq \mathbf{P}(D_{k-1}) + \mathbf{P}(F_{k-1}^{(1)})\alpha + \mathbf{P}(F_{k-1}^{(2)})\alpha + \cdots + \mathbf{P}(F_{k-1}^{(v)})\alpha$$

$$= \mathbf{P}(D_{k-1}) + [\mathbf{P}(F_{k-1}^{(1)}) + \mathbf{P}(F_{k-1}^{(2)}) + \cdots + \mathbf{P}(F_{k-1}^{(v)})]\alpha. \tag{12}$$

But the sum of the probabilities of the pairwise mutually exclusive events $D_{k-1}, F_{k-1}^{(1)}, F_{k-1}^{(2)}, \ldots, F_{k-1}^{(v)}$ is equal to 1, since these events form a complete system (Formula (7) of section 3). Hence, the sum in the square brackets is equal to

$$1 - \mathbf{P}(D_{k-1}),$$

and Formula (12) takes the form

$$\mathbf{P}(D_k) \geq \mathbf{P}(D_{k-1}) + [1 - \mathbf{P}(D_{k-1})]\alpha. \tag{13}$$

By Property 3,

$$\mathbf{P}(\overline{D}_k) = 1 - \mathbf{P}(D_k),$$

$$\mathbf{P}(\overline{D}_{k-1}) = 1 - \mathbf{P}(D_{k-1}).$$

From this and (13), we obtain

$$\mathbf{P}(\overline{D}_k) = 1 - \mathbf{P}(D_k) \leq 1 - \mathbf{P}(D_{k-1}) - [1 - \mathbf{P}(D_{k-1})]\alpha$$

$$= \mathbf{P}(\overline{D}_{k-1}) - \mathbf{P}(\overline{D}_{k-1})\alpha = \mathbf{P}(\overline{D}_{k-1})(1 - \alpha).$$

Proof. Part 2. We now seek numbers N and α possessing the desired properties.

We choose an arbitrary intersection E_i. The traveler can reach the intersection E_0 from the intersection E_i by moving along various paths. We choose one of the paths $E_iE_jE_k \ldots E_rE_sE_0$. We denote the number of moves in this path by N_i. Let us calculate the probability α_i that the traveler traverses the path $E_iE_jE_k \ldots E_rE_sE_0$ if he starts at E_i.

We denote the probability that the traveler goes from the intersection E_i along the path E_iE_j by p_{ij}. The probability p_{ij} is the conditional probability that the traveler is at E_j after the nth move, under the condition that he was at E_i after the $(n-1)$st move. Then the probability that the traveler starting out from E_i takes the path $E_iE_jE_k$ (Fig. 14) is equal to $p_{ij}p_{jk}$. Indeed, by Property 6,

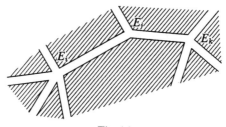

Fig. 14

this probability is the product of the probability that the traveler is at E_j after one move and the probability (under this condition) that he is at E_k after one further move. The first factor is equal to p_{ij}, the second is equal to p_{jk}; hence, the desired probability is equal to $p_{ij}p_{jk}$. In general, the probability that a traveler starting out from E_i chooses the path $E_iE_jE_k \ldots E_rE_sE_0$ is $p_{ij}p_{jk} \ldots p_{rs}p_{s0} = \alpha_i$.

We carry through this calculation for every intersection E_1, E_2, $\ldots E_v$. We thus obtain the numbers $N_1, N_2, \ldots, N_v$; $\alpha_1, \alpha_2, \ldots, \alpha_v$. Let N be the greatest of the numbers $N_1, N_2, \ldots, N_v$, and α be the smallest of the numbers $\alpha_1, \alpha_2, \ldots, \alpha_v$. We must show that N and α have the desired properties.

First of all, the numbers $\alpha_1, \alpha_2, \ldots, \alpha_v$ are all positive; hence, we also have $\alpha > 0$.

Suppose the traveler has traveled N moves from the intersection E_i. He has thereby reached the point E_0 at least once, with a probability greater than or equal to α; the event "the traveler took the path $E_iE_jE_k \ldots E_rE_sE_0$ in the first N_i moves" implies the event "the traveler reached E_0 at least once in N moves" (for if the first event occurs, the traveler reaches E_0 after having made N_i moves). By Property 1, the probability of the second event is greater than or equal to the probability of the first, which is $\alpha_i \geq \alpha$.

Proof. Part 3. We estimate, successively, the probabilities $\mathbf{P}(\bar{D}_1)$, $\mathbf{P}(\bar{D}_2), \ldots, \mathbf{P}(\bar{D}_k), \ldots$.

First of all, $\mathbf{P}(D_1) \geq \alpha$, and hence, by Property 3,

$$\mathbf{P}(\bar{D}_1) = 1 - \mathbf{P}(D_1) \leq 1 - \alpha.$$

Furthermore, we find, with the help of the formula

$$\mathbf{P}(\bar{D}_k) \leq \mathbf{P}(\bar{D}_{k-1})(1 - \alpha),$$

obtained in the first part of the proof, that

$$\left.\begin{aligned}
\mathbf{P}(\bar{D}_2) &\leq \mathbf{P}(\bar{D}_1)(1 - \alpha) \leq (1 - \alpha)^2, \\
\mathbf{P}(\bar{D}_3) &\leq \mathbf{P}(\bar{D}_2)(1 - \alpha) \leq (1 - \alpha)^3, \\
&\cdots\cdots\cdots\cdots\cdots\cdots\cdots\cdots \\
\mathbf{P}(\bar{D}_k) &\leq \mathbf{P}(\bar{D}_{k-1})(1 - \alpha) \leq (1 - \alpha)^k, \\
&\cdots\cdots\cdots\cdots\cdots\cdots\cdots\cdots
\end{aligned}\right\} \quad (14)$$

We consider the event $\bar{D}$ (that the traveler never reaches E_0). $\bar{D}$ implies each of the events $\bar{D}_1, \bar{D}_2, \ldots, \bar{D}_k, \ldots$. Hence, $\mathbf{P}(\bar{D})$ is, by Property 1, not greater than the probabilities $\mathbf{P}(\bar{D}_1), \mathbf{P}(\bar{D}_2), \ldots,$ $\mathbf{P}(\bar{D}_k), \ldots$ and, hence, not greater than any of the numbers

$$1 - \alpha, (1 - \alpha)^2, \ldots, (1 - \alpha)^k, \ldots \quad (15)$$

Since $\alpha > 0$ and $1 - \alpha < 1$, the terms of the sequence (15) form an infinite decreasing geometric progression, which for increasing k becomes smaller than any arbitrary positive number ε. Hence, the probability that E_0 is never reached is smaller than any positive number, and is thus equal to zero. By Property 3, the probability that E_0 will be reached is then equal to 1.

The following theorem has thus been proved:

THEOREM. *The traveler reaches every intersection with a probability of* 1 *regardless of the intersection from which he starts.*

Remark 1. We consider two arbitrary intersections, denote them by E_0 and E_1, and assume that the traveler starts his path at the intersection E_1. Let us calculate the probability of the events that

A_0 : The traveler reaches E_0 at least once;

A_{01} : The traveler reaches E_0 and then the intersection E_1;

A_{010}: The traveler reaches E_0, then reaches E_1, and after that returns to E_0, etc.

By the theorem just proved, we have $\mathbf{P}(A_0) = 1$. By Property 6, the probability of the event A_{01} is equal to the product of $\mathbf{P}(A_0)$ and the probability that the traveler, on going out from E_0, then reaches E_1. By our theorem, both factors are equal to 1; hence, $\mathbf{P}(A_{01}) = 1$. Furthermore, the probability $\mathbf{P}(A_{010})$ is, by the same Property 6, equal to the product of $\mathbf{P}(A_{01})$ and the probability that the traveler, on going out from E_1, finally reaches E_0. From this, it is obvious that

$$\mathbf{P}(A_{010}) = 1.$$

In a similar manner, we show that

$$\mathbf{P}(A_{0101}) = \mathbf{P}(A_{01010}) = \cdots = 1.$$

We pick an arbitrary positive integer k. It follows from what has been proved that:

a) With the probability 1, the traveler returns to the starting point at least k times.

b) With the probability 1, the traveler reaches an arbitrary preassigned intersection E_0 at least k times.

Remark 2. The formula

$$\mathbf{P}(\bar{D}_k) \leq (1 - \alpha)^k$$

(compare with (14)) gives an estimate for the probability of error of the statement that after kN moves the traveler has reached E_0 at least once. We choose an arbitrary probability ε. Then, we can find a number k such that

$$(1 - \alpha)^k < \varepsilon.$$

But then, certainly,

$$\mathbf{P}(\bar{D}_k) < \varepsilon,$$

and the statement that after kN moves the traveler has reached E_0 at least once, has a probability of error less than ε. Thus, even for arbitrarily great demands on the degree of certainty, one can give a number of moves in which the traveler reaches E_0 with practical certainty.

Formula (14) is general; that is, it permits an approximation of the probability $\mathbf{P}(\bar{D}_k)$ for an arbitrary real city. However, it yields only a rough approximation to this probability. We shall apply this approximation to the consideration of an earlier example (see Fig. 13). In this example, one can take $N = 3$ and $\alpha = \frac{1}{4}$. Hence, the probability that after $k \cdot 3$ moves the traveler never reaches the point 3 satisfies the inequality

$$\mathbf{P}(\overline{D}_k) \leq \left(\frac{3}{4}\right)^k.$$

We ask that

$$\left(\frac{3}{4}\right)^k < 0.01.$$

The smallest value of k that satisfies this inequality is equal to 17 (see p. 39). Therefore, the smallest suitable number of subsequences is equal to 17, and we reach the point 3 after $17 \cdot 3 = 51$ moves with a probability of at least 0.99. However, as was shown earlier, even 34 moves are sufficient (not less). Thus, even in this simple case, our approximation yields a result which is less accurate than the exact calculation by a factor of one and one half. In more complicated examples, the result becomes still less precise.

Our traveler need not go on foot; he can use one of the municipal methods of conveyance: streetcar, bus, trolley bus, or subway. At a stop there is a certain probability that he enters, say, a bus. After that, he gets off with a certain probability at each of the following stops and continues with a certain probability.

In this case, the random path of the traveler through the city does not differ in essence from the board game *Circus* mentioned in the Introduction. One can instead name the game *Journey Through a City*. The squares of the board are represented as street intersections, and municipal conveyances substituted for the circus attractions. The sudden motion to another edge of the board corresponds, perhaps, to the journey to the next subway station. We wish to use a device of chance that offers more possibilities than a die. One can, for example, use an urn, and determine the direction of motion by drawing from the urn a piece of paper bearing the designation of an intersection. It is clear, however, that a single urn will not suffice, since the point that we reach depends on the point from which we start. Ideally, one would use as many different urns as there are squares on the board.

12. MARKOV CHAINS

We now consider an arbitrary diagram of points $E_1, E_2, \ldots, E_n$, several of which are joined by arrows that point in the possible directions of motion.

DEFINITION. *A system of n points or states for which we know the possibilities of transitions between them as well as the probabilities that these transitions take place, is called a* Markov chain.

The probability that one goes in one step from E_i to E_j is generally denoted by p_{ij} (in particular, p_{ii} denotes the probability of not moving from E_i on a move).[1] The transition from E_i to E_j is possible if $p_{ij} > 0$ (in this case, we draw an arrow from E_i to E_j).

DEFINITION. *A* Markov chain *is called* irreducible *if one can go from any position E_i to any other position E_j by means of a chain of possible transitions.*

In terms of the arrows, this means that one can go from any point E_i to any other point E_j in the direction of the arrows. Figure 15 gives an example of a chain that is not *irreducible* (it is impossible to go from E_2 to E_1 in the direction of the arrows). Figures 16 to

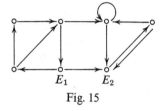

Fig. 15

20 are examples of irreducible chains. We have already considered a *Markov chain* of two states in Problem 5 (Fig. 4).

The following general theorem holds:

THEOREM. *In the motion of a particle through an arbitrary system of states that form an irreducible Markov chain, this particle reaches any state with a probability of 1 (independent of its starting point).*

This theorem was already proved, in essence, as a theorem on the random walk of a traveler through a city. The reader can ascertain without difficulty that our proof is also valid, step for step, without an alteration for arbitrary irreducible Markov chains.

The Markov chains are named after the noted Russian mathematician Andrei Andreievich Markov (1856–1922), who discovered

---

[1] We remark that, for every i,
$$p_{i1} + p_{i2} + \cdots + p_{ii} + \cdots + p_{in} = 1,$$
since $p_{i1}, p_{i2}, \ldots, p_{in}$ are the probabilities of mutually exclusive events (going from E_i to E_1, from E_i to E_2, and finally from E_i to E_n) that form a complete system.

and investigated them. Markov chains are very important because of their applications to science and technology, as well as for probability theory itself.

In the solution of Problem 5 it was remarked that for a chain of two states the probability that after n moves a given position is reached depends less and less on the starting point with increasing n. This fact holds for all irreducible Markov chains (with minor limitations).

13. THE MEETING PROBLEM

We return once more to the city plan in Figure 1. Suppose that, as before, the friends start out on their walk from intersection A, but that this time they go their own ways; that is, each of them tosses two coins independently of the other and chooses his path according to the outcome of his own toss. Will the friends meet again after they have left A? We shall show that with probability 1 this meeting will take place somewhere. Moreover, if any crossing E is fixed beforehand, it can even be asserted that with probability 1 the friends will meet at E.

We restate this problem in general terms as follows:

THEOREM. *Two particles move along an irreducible chain K, beginning their motion at the same time at an arbitrary state. At each move, each particle moves independently of the other, from one state to another. The probability that the two particles meet in an arbitrary preassigned state is equal to 1.*

First, we explain an important relation between Markov chains.

Fig. 16

Let the chain K consist of n states E_1, E_2, ..., E_n. We consider n^2 points, and denote them by E_{11}, E_{12}, ..., E_{1n}, E_{21}, E_{22}, ..., E_{2n}, ..., E_{n1}, E_{n2}, ..., E_{nn}. We shall denote the state of the pair of particles on the chain K by a marker that is situated on one of the points E_{11}, ..., E_{nn}. If the first particle is in state E_i, and the second is in state E_k, we place the marker on the point E_{ik}. The passage of the first particle from E_i to E_j occurs with the probability p_{ij}, and the passage of the other particle from E_k to E_l

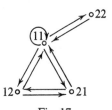

Fig. 17

with the probability p_{kl}. The simultaneous passage of the first particle from E_i to E_j, and the second from E_k to E_l occurs with the probability $p_{ij}p_{kl}$ (by Formula (4), section 2), and, hence, the marker goes from E_{ik} to E_{jl} with the probability $p_{ij}p_{kl}$. We have thus obtained a new Markov chain, which we denote by K^2. The chain K^2 obtained from the chain K shown in Figure 16 can be seen in Figure 17.

According to the general rule, we can draw an arrow from E_{ik} to E_{jl} when the probability of the passage from E_{ik} to E_{jl} is positive. From this it follows that there is an arrow from E_{ik} to E_{jl} if and only if the two probabilities p_{ij} and p_{kl} are positive, that is, if and only if an arrow leads from E_i to E_j and an arrow leads from E_k to E_l. The system of arrows in the chain K^2 is thus formed from the system of arrows in the chain K, the value of p_{ij} being unimportant.

Problem 21. (*a*) The system of arrows of a chain K is shown in Figure 18. Construct the system for the chain K^2;

    (*b*) Do the same for the system shown in Figure 19;

    (*c*) Do the same for the system shown in Figure 20.

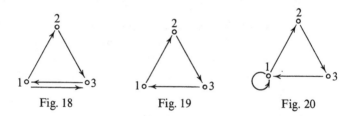

Fig. 18   Fig. 19   Fig. 20

We consider the totality L of states in K^2 which one can reach from E_{11} by moving along the arrows.

Problem 22. Prove that:

    (*a*) The positions E_{22}, E_{33}, ..., E_{nn} are contained in L.

    (*b*) L is an irreducible chain.

The proof of the meeting theorem can now be given in a few words.

Proof. Let K be an arbitrary irreducible chain, and suppose that the two particles begin their motion at the same time in the state E_{11}. We construct the chain K^2. This may not be irreducible (as was shown, for example, in Problem 21*b*). We form the chain L from K^2. This chain is (as was shown in Problem 22*b*) irreducible, and we can apply the general theorem formulated on page 46. Hence, after leaving the state E_{11}, the marker will reach any state of L with probability 1, including E_{11}, E_{22}, ..., E_{nn}. But this means that our particles meet with probability 1 in any arbitrary preassigned state.

It is proved in exactly the same way that 3 or 4 or, in general, n particles that begin their motion at the same time at a state of an irreducible chain, meet again with probability 1, and furthermore, with probability 1 they reach any arbitrary preassigned state at the same time.

4. Random Walks with Infinitely Many States

The Markov chains that we considered in the previous chapter had finitely many states. We now turn to a consideration of chains with infinitely many states. (We have already met one such chain in the diagram of the random walk on a line.) Exactly as in the previous section, the question of interest to us is whether a particle reaches a given point, and if it does, how fast. The random walk of a particle on an infinite chain differs qualitatively from the motion on a finite chain. For example, one cannot in general assert here that the particle reaches every position with the probability 1 (although this is possible in individual cases).

We limit ourselves to the simplest chains with infinitely many states. First, we investigate the scheme, already familiar to us, of the random walk on a straight line (we shall also call this chain an "infinite path"). Then, we shall consider another simple example of a chain with infinitely many states, an infinitely large city with a checkerboard pattern (Fig. 21). If a traveler reaches any crossing, he continues in any of the four directions with a probability $\frac{1}{4}$.

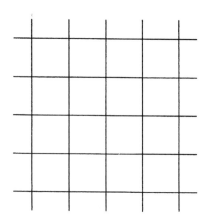

Fig. 21

14. RANDOM WALKS ON AN INFINITE PATH

Problem 23. Let the points $0, \pm 1, \pm 2, \pm 3, \ldots$ be marked on a line (Fig. 22). A particle located at a point n might move at the next

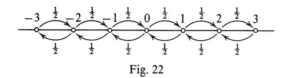

Fig. 22

moment to the point $n + 1$ with the probability $\frac{1}{2}$ and to the point $n - 1$ with the same probability. At the beginning, the particle is situated at the point 0. Find:

(*a*) The probability x that the particle reaches the point 1 at least once.

(*b*) The probability y that the particle reaches the point -1 at least once.

(*c*) The probability z that the particle returns to the point 0 at some time (that is, that it is situated at the zero point at a time other than the beginning of the random walk).

Problem 24. Prove that the particle of Problem 23 reaches every point with the probability 1.

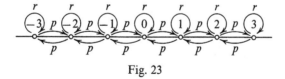

Fig. 23

Problem 25. Suppose that in the diagram in Figure 23 the particle moves from the point n to the point $n + 1$ in a unit of time with probability p, moves to the point $n - 1$ with the same probability p, and remains in the same place with the probability r (in Problem 23, $p = \frac{1}{2}$ and $r = 0$). Prove that the particle reaches every point with the probability 1 for $p > 0$, regardless of the starting point.

It is established for the chains in Problems 24 to 27 that the particle, wherever it might start, reaches every position with probability 1. However, the chain shown in Figure 24 does not have

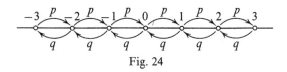

Fig. 24

this property; here, the particle goes from n to $n + 1$ with the probability p, and from n to $n - 1$ with the probability $q = 1 - p$, where $p > q$. One can prove (we shall refrain from doing so here) that the probability that the particle ever reaches the point -1 after starting out from the point 0 is less than 1.

Problem 26. Using the fact that in Figure 24 the probability of ever reaching the point -1 from the point 0 is smaller than 1, find the probabilities x, y, z (of Problem 23).

In Problem 23, it developed that a particle moving on the line returns to its starting point with a probability 1. How many moves must it make to make this return practically certain? The calculation shows that, for a probability greater than 0.99, several thousand moves are necessary; if this probability is to be greater than 0.999, then several hundred thousand moves are necessary.

One can show that the probability of a return is less than $1 - \varepsilon$ when the particle has made less than $\dfrac{0.2}{\varepsilon^2}$ moves. When it has made more than $\dfrac{0.87}{\varepsilon^2}$ moves, this probability is greater than $1 - \varepsilon$.

It was further shown that the particle reaches every point with probability 1. We ask again, how many moves are necessary for the particle to have reached a given point with practical certainty. It is clear that the farther the point of interest to us is from the starting point, the greater the number of moves necessary for the particle to reach it. To reach the point 10 with a probability of 0.99, more than 100,000 moves are necessary; to reach the point 100 with the same probability, tens of millions of moves are necessary.

We set

$$N_1 = \frac{2(\frac{3}{4}k - 1)^2}{\varepsilon^2},$$

$$N_2 = \frac{10k^2}{\varepsilon^2} + \frac{6}{\varepsilon} + 2.$$

One can prove that the probability that the particle reaches the point $2k$ even once is smaller than $1 - \varepsilon$, if it makes less than N_1 moves.[1] If it makes more than N_2 moves, the probability of the same event is greater than $1 - \varepsilon$.

15. THE MEETING PROBLEM

We now turn to the problem of determining the probability that two travelers on an infinite path K will meet (Fig. 22). In the previous chapter, we solved the analogous problem for an arbitrary Markov chain with finitely many states by reducing it to the problem of a single particle moving on a new, more complicated Markov chain with finitely many states. This device was sufficient for the solution of the meeting problem, since we had a general theorem about the random motion of particles that held for all arbitrarily complicated irreducible Markov chains with finitely many states. In the case of infinitely many states, the meeting problem cannot be solved in this fashion, since we lack a corresponding theorem holding for all chains with infinitely many states. We are therefore compelled to solve the meeting problem by a direct method.

In the case of infinitely many states, we cannot reduce the motion of two particles to the motion of one particle on a more complicated chain. On the contrary, we must reduce the problem of motion of a particle on a complicated chain to a problem on the random motion of two (or several) particles on a simpler chain. We shall proceed in this way in the following investigation of an infinitely large city with a checkerboard pattern. We shall reduce the motion of a particle (marker) in this city to the motion of two particles (travelers) on an infinite path K.

We make use of the following auxiliary problem.

---

[1] This statement holds for $k \geq 2$, $\varepsilon \leq \frac{3}{4} - \frac{1}{k}$.

Problem 27. Prove that the sum

$$S_k = \frac{1}{1} + \frac{1}{2} + \frac{1}{3} + \frac{1}{4} + \cdots + \frac{1}{k-1} + \frac{1}{k}$$

increases with k and can surpass any arbitrary value.

THEOREM. *Suppose the travelers begin their journey at the same time at point 0. Then they meet again at this point with probability 1.*

Proof. Part 1. We consider the events:

A_s: Both travelers reach the point 0 after s moves.

B_s: Both travelers reach the point 0 after s moves, without a previous meeting[1] at this point.

C_s: Up to and including the sth move, the travelers have not met at the point 0.

The events $B_1, B_2, B_3, \ldots, B_{s-1}, B_s, C_s$ are pairwise mutually exclusive and form a complete system. Hence (Property 7),

$$\mathbf{P}(A_s) = \mathbf{P}(B_1)\,\mathbf{P}(A_s|B_1) + \mathbf{P}(B_2)\,\mathbf{P}(A_s|B_2) + \cdots$$
$$+ \mathbf{P}(B_s)\,\mathbf{P}(A_s|B_s) + \mathbf{P}(C_s)\,\mathbf{P}(A_s|C_s). \qquad (1)$$

Clearly,

$$\mathbf{P}(A_s|C_s) = 0, \ \mathbf{P}(A_s|B_s) = 1.$$

Furthermore, $\mathbf{P}(A_s|B_i)$ is the probability that both travelers are at point 0 after the sth move, under the condition that they had already met there after the ith move; that is, $\mathbf{P}(A_s|B_i)$ is the probability that the two travelers, starting out from 0, return there after $s - i$ moves. Hence,

$$\mathbf{P}(A_s|B_i) = \mathbf{P}(A_{s-i}).$$

If we now set

$$\mathbf{P}(A_s) = a_s,$$

and

$$\mathbf{P}(B_s) = b_s,$$

equality (1) assumes the form

$$a_s = b_1 a_{s-1} + b_2 a_{s-2} + \cdots + b_{s-1} a_1 + b_s.$$

[1] Except, of course, at the initial moment.

We write this equality out for $s = 1, 2, 3, \ldots, n$ and add (we write out those a_k, b_k that are equal to zero for the sake of clarity):

$$
\left.
\begin{aligned}
a_1 &= b_1, \\
a_2 &= b_2 + a_1 b_1, \\
a_3 &= b_3 + a_1 b_2 + a_2 b_1, \\
&\cdots\cdots\cdots\cdots\cdots\cdots\cdots\cdots\cdots\cdots\cdots\cdots\cdots \\
a_n &= b_n + a_1 b_{n-1} + a_2 b_{n-2} + \cdots + a_{n-1} b_1 \\
\hline
R_n &= Q_n + a_1 Q_{n-1} + a_2 Q_{n-2} + \cdots + a_{n-1} Q_1 .
\end{aligned}
\right\} \quad (2)
$$

Here we have the set

$$
R_n = a_1 + a_2 + a_3 + \cdots + a_n,
$$
$$
Q_n = b_1 + b_2 + b_3 + \cdots + b_n.
$$

Clearly, Q_n is the probability of the event D_n: "in the course of the first n moves, the travelers meet at least once at the point 0"; for we have

$$
D_n = B_1 + B_2 + B_3 + \cdots + B_n,
$$
$$
\mathbf{P}(D_n) = \mathbf{P}(B_1) + \mathbf{P}(B_2) + \cdots + \mathbf{P}(B_n)
$$
$$
= b_1 + b_2 + \cdots + b_n
$$
$$
= Q_n.
$$

The probability Q that the travelers meet at 0 at some time (the probability we are seeking) is, by Property 1, greater than or equal to Q_n; that is,

$$
Q \geq Q_n, \quad (3)
$$

for every n.

It follows from (2) and (3) that

$$
R_n \leq Q + a_1 Q + a_2 Q + \cdots + a_{n-1} Q = (1 + R_{n-1})Q,
$$
$$
Q \geq \frac{R_n}{1 + R_{n-1}}. \quad (4)
$$

The inequality (4) holds for every n.

Proof. Part 2. We shall now show that R_n exceeds every arbitrary number with increasing n; we have $R_n = a_1 + a_2 + a_3 + \cdots + a_n$, where a_n is the probability that the two travelers meet at 0 after n moves. Since the two travelers move independently of each other, a_n can be found by the multiplication formula for probabilities:

$$a_n = w_n{}^2,$$

where w_n is the probability that one traveler is at point 0 after n moves. The probability w_{2k+1} is, obviously, equal to zero. As was shown in Chapter 2,

$$w_{2k} \geq \frac{1}{\sqrt{4k}}.$$

Hence,

$$a_{2k+1} = 0, \quad a_{2k} \geq \frac{1}{4k}.$$

From this it follows that

$$R_{2k+1} = R_{2k} \geq \frac{1}{4} S_k,$$

where $S_k = 1 + \frac{1}{2} + \frac{1}{3} + \cdots + \frac{1}{k}$. Since, however, S_k increases without bound, R_n also increases without bound.

Proof. Part 3. We now prove that $Q = 1$. First of all, Q, like all probabilities, is either less than or equal to 1. We assume that $Q = 1 - d$, where $d > 0$, and come to a contradiction. By inequality (4),

$$1 - d \geq \frac{R_n}{1 + R_{n-1}} = \frac{R_{n-1} + a_n}{R_{n-1} + 1},$$

$$R_{n-1} + 1 - d(R_{n-1} + 1) \geq R_{n-1} + a_n,$$

$$1 - a_n \geq d(R_{n-1} + 1),$$

$$1 > dR_{n-1},$$

$$R_{n-1} < \frac{1}{d}.$$

But this is false, since R_{n-1} increases without bound for increasing n.

16. THE INFINITELY LARGE CITY WITH A CHECKERBOARD PATTERN

We have already said that we shall reduce the motion through an infinitely large city with a checkerboard pattern to the motion of two travelers on an infinite path K.

Let us imagine that two travelers move on the path K. The chain K^2 is constructed by the process given in the preceding chapter, and consists of the positions $(0, 0)$, $(0, 1)$, $(0, -1)$, $(1, 1)$, $(1, -1)$, $(-1, 1)$, $(-1, -1)$, etc. The chain K^2 is shown in Figure 25.

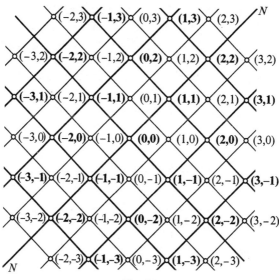

Fig. 25

Suppose that the travelers are at the points m and n of the chain K. The first traveler goes with a probability $\frac{1}{2}$ from m to $m + 1$, and the second traveler goes with the same probability from n to $n + 1$. Hence, the probability that the two travelers go from the point pair (m, n) to the point pair $(m + 1, n + 1)$ is equal to $\frac{1}{2} \cdot \frac{1}{2} = \frac{1}{4}$. This means that the marker that moves on the chain K^2 and reflects the position of both travelers on K goes from (m, n) to $(m + 1, n + 1)$ with the probability $\frac{1}{4}$. The marker goes from (m, n) to each of the other points

$$(m + 1, n - 1), (m - 1, n + 1), (m - 1, n - 1)$$

neighboring (m, n) with the same probability $\frac{1}{4}$.

We remark that at each move the distance between the travelers either changes by two or not at all. Therefore, if at the beginning the travelers are at the points k and l, they can subsequently reach at the same time only such points m and n whose difference $m - n$ is of the same parity as $k - l$. Hence, the chain K^2 is not irreducible; it divides into two irreducible chains: into L, containing the pairs (m, n) with even difference (that is, the points that in Figure 25 are joined by thick lines), and M, containing the pairs with odd difference (that is, the points in Figure 25 that are joined by thin lines).

We assume that the marker starts out from the position (0, 0) of the chain L (corresponding to the fact that both travelers start out from the position 0 of the chain K at the same time). The marker must remain on the boundaries of L in its further motion. The chain L is shown separately in Figure 26, where it is rotated 45° from its

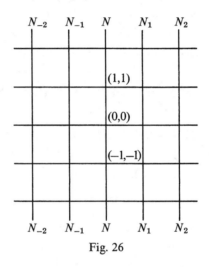

Fig. 26

position in Figure 25. However, Figure 26 coincides with Figure 21. We find, therefore, that the problem of the random walk of two travelers on the path K is equivalent to the problem of the random walk of one traveler through an infinitely large city of checkerboard pattern.

We have proved that two travelers starting out from the point 0 of the path K at the same time, will, with probability 1, again meet at this point. In terms of the city L, this means that a marker that begins its motion at (0, 0) returns there with a probability 1. Thus, simply translating the result of the previous section into this new language, we have proved that *the marker returns to the starting point with probability 1.*

The return of the marker to the starting point takes much longer in the infinitely large city with a checkerboard pattern than on the infinite path; that is, many more moves must be made for this return to be practically certain.

For example, to obtain a probability of return of more than 0.99 the marker must make an astronomical number of moves: more than 10^{88} (in the case of the path K, 10,000 moves sufficed).

In general, if the marker makes less than $10^{\frac{0.9}{\varepsilon}-1.7}$ moves, the probability that it returns at least once to the starting point is less than $1 - \varepsilon$. This probability becomes greater than $1 - \varepsilon$ when the marker makes more than $10^{\frac{2}{\varepsilon}-1.6}$ moves.

We now prove that the marker moving about the infinitely large city with a checkerboard pattern not only returns to the starting point with probability 1, but also reaches any arbitrary preassigned intersection with probability 1.

Let E_0 and E_1 be two neighboring intersections. By x, we denote the probability that the figure reaches the point E_1 at any time, having started out from E_0. Due to the uniform construction of our city, this probability has the same value for any pair of neighboring intersections.

Suppose that the marker starts out from E_0. After the first move it can have reached one of the four intersections $0_1, 0_2, 0_3, 0_4$ next to E_0. We consider the events:

A_i: After one move, the marker reaches the point 0_i.
B: The marker returns at any time to E_0.

By Chapter 1, Property 7,

$$\mathbf{P}(B) = \mathbf{P}(A_1)\mathbf{P}(B|A_1) + \mathbf{P}(A_2)\mathbf{P}(B|A_2)$$
$$+ \mathbf{P}(A_3)\mathbf{P}(B|A_3) + \mathbf{P}(A_4)\mathbf{P}(B|A_4). \qquad (5)$$

From the plan of the city, one sees that

$$\mathbf{P}(A_1) = \mathbf{P}(A_2) = \mathbf{P}(A_3) = \mathbf{P}(A_4) = \frac{1}{4}.$$

The occurrence of B under the condition A_i means that the marker reaches the point E_0 on starting out from the point 0_i (to which it came after the first move).

Hence,

$$\mathbf{P}(B|A_1) = \mathbf{P}(B|A_2) = \mathbf{P}(B|A_3) = \mathbf{P}(B|A_4) = x.$$

Finally, as was proved above, $\mathbf{P}(B) = 1$. If all these values are substituted in Formula (5), we obtain

$$1 = \frac{1}{4}x + \frac{1}{4}x + \frac{1}{4}x + \frac{1}{4}x,$$

and thus, $x = 1$.

Now, let E be an arbitrary intersection of the checkerboard city. We consider one of the paths that lead from E_0 to E, and number in order all of the intersections that belong to this path: E_0, $E_1, E_2, \ldots, E_n = E$.

By D we mean the event that the marker, on starting out from E_0, reaches E_1, then E_2, etc., and finally, $E_n = E$.

The probability that the marker reaches E_{i+1} at some time after starting out from E_i is, as was calculated, equal to 1. If we multiply these probabilities for $i = 0, 1, 2, \ldots, n - 1$ according to the multiplication formula for probabilities, we obtain $\mathbf{P}(D)$. Clearly, the occurrence of the event D implies the occurrence of the event C, that the marker reaches E at some time.

By Chapter 1, Property 1,

$$\mathbf{P}(C) \geq \mathbf{P}(D) = 1,$$

and, consequently, $\mathbf{P}(C) = 1$.

The consideration of the random walk through a city of the kind given makes it possible to sharpen still more the formulation dealing with the infinite path. It was shown that the travelers moving on the infinite path meet at the starting point with a probability 1. We can now assert that the travelers meet at any arbitrary preassigned point n with the probability 1. For this is equivalent to the statement that a marker moving in the city with a checkerboard pattern reaches the intersection (n, n) with probability 1.

This result, however, cannot be generalized to arbitrarily many travelers as in the case of a finite Markov chain. For four or more travelers moving on the path K, the probability that they all meet is smaller than 1. For three travelers, the probability that they all meet somewhere is equal to 1, but the probability that they meet in a preassigned point is less than 1.

We can summarize the results of our investigations of the problems on motion and meeting in a city of checkerboard design in the following theorems:

THEOREM 1. *Let A and B be two arbitrary points on the path K. Two travelers that start out from the point A at the same time meet at the point B with probability* 1.

THEOREM 2. *Let A and B be two arbitrary intersections in the city L. A traveler starting from the intersection A reaches the intersection B with probability* 1.

The proofs of these theorems are intimately connected. The proof of the second theorem arises from the special case of the first, in which the points A and B coincide. The general formulation of Theorem 1 arises, in turn, from Theorem 2.

The special case of Theorem 1 that constitutes the first link in our chain of reasoning demands for its proof some calculations. These calculations can be avoided if one contents himself with a weaker assertion than that of Theorem 1.

THEOREM 1*a*. *Two travelers starting out at the same time from an arbitrary point n of the path K meet again with probability* 1.

If we carry over this statement to the context of our city, it assumes the following form:

THEOREM 2*a*. *A traveler starting out from the intersection (n, n) of the city L reaches, with probability* 1, *a point of the line NN* (Fig. 26).

The proof of this last theorem can be carried through with the aid of Problem 25 alone. Hence, we pose the following problem:

Problem 28. Deduce Theorem 2*a* with the aid of Problem 25.

Concluding Remarks

In this booklet we have considered problems in probability theory. It was our aim to acquaint the reader with the concepts and methods of this unique science by means of examples that were sufficiently clear and which, at the same time, required for their consideration more complicated processes than the simple calculation of the number of desired outcomes. The fact is that the overwhelming majority of the problems that are dealt with by modern probability theory cannot be solved by such simple calculations.

On the other hand, we could not consider more complicated and interesting examples, since they require for their solution tools that lie outside the realm of elementary mathematics. However, we do not wish to leave the reader with the impression that probability theory is the science of children's games and strolls through a city. In reality, the domain of application of probability theory is very great. Probability theory is widely used in technology (radio engineering, the design of telephone networks, quality control of production, etc.), ballistics (the investigation of the scattering of shots), the evaluation of experimental results (the theory of errors). Probability theory also finds important and diversified applications in physics. We have already spoken of some of them (Brownian motion).

Beginning with the work of the great mathematician P. L. Chebyshev (1821–1894), Russian scientists have taken a leading role in the theory of probability. Chebyshev's work has been continued by his students, A. A. Markov (1856–1922) and A. M. Lyapunov (1857–1918). The Soviet school of probability theory has produced such distinguished scholars as S. N. Bernstein (1880–   ), A. N. Kolmogorov (1903–   ), and A. Ya. Khinchin (1894–1959).

The following literature is recommended:

Feller, William. *An Introduction to Probability Theory and Its Applications*. Vol. I. 2d ed. (New York: John Wiley & Sons, Inc., 1957.)

Gnedenko, B. V., and Khinchin, A. Ya. *Elementary Introduction to the Theory of Probability*. Translated by W. R. Stahl. (San Francisco and London: W. H. Freeman and Company, 1961.)

Goldberg, Samuel. *Probability, an Introduction*. (Englewood Cliffs, N.J.: Prentice-Hall, Inc., 1960.)

Wolf, Frank L. *Elements of Probability and Statistics*. (New York: McGraw-Hill Book Company, Inc., 1962.)

Solutions to Problems

PROBLEM 1. There are 7 possible events, of which 3 satisfy the condition. Hence, the desired conditional probability is equal to $\frac{3}{7}$.

PROBLEM 2. The probability that a six does not come up on one toss is equal to $\frac{5}{6}$. The probability that no six appears on six tosses is, by formula (6) on page 10,

$$\frac{5}{6} \cdot \frac{5}{6} \cdot \frac{5}{6} \cdot \frac{5}{6} \cdot \frac{5}{6} \cdot \frac{5}{6} = \left(\frac{5}{6}\right)^6 = \frac{15{,}625}{46{,}656} \approx 0.34.$$

PROBLEM 3. We denote the desired probability by x. Then, the probability of no hits in n shots is equal to $1 - x$, of no hits on one shot is equal to $1 - p$ (Property 3). Exactly as in the previous exercise we have, by formula (6),

$$1 - x = (1 - p)^n,$$

and thus,

$$x = 1 - (1 - p)^n.$$

PROBLEM 4. We denote the probability of the event A, that the first player wins, by x, the probability of the event B, that the second player wins, by y, and the probability of the event C, that the game never ends, by z.

The events A, B, and C are pairwise mutually exclusive and form a complete system; hence,

$$x + y + z = \mathbf{P}(A) + \mathbf{P}(B) + \mathbf{P}(C) = 1. \tag{1}$$

The probability that the second player wins is equal to the probability that tails comes up on the first toss (this probability is equal to $\frac{1}{2}$) multiplied by the probability that this player wins under this condition (Property 6). Then, however, the second player is in the position of the first, and, hence, this conditional probability is equal to x. Thus,

$$y = \frac{1}{2}x. \tag{2}$$

The probability that the first player wins is found by the formula for the complete probability. We consider the events:

D: "Heads comes up on the first toss."
F: "Tails comes up on the first toss."

The events D and F are mutually exclusive and form a complete system; by Property 7, we have

$$x = \mathbf{P}(A) = \mathbf{P}(D)\,\mathbf{P}(A|D) + \mathbf{P}(F)\,\mathbf{P}(A|F).$$

Obviously, $\mathbf{P}(D) = \mathbf{P}(F) = \frac{1}{2}$ and $\mathbf{P}(A|D) = 1$. To calculate $\mathbf{P}(A|F)$, we remark that the first player is in the position of the second, and the second player in the position of the first, when tails comes up on the first toss; the probability of his winning is then y. Hence,

$$x = \frac{1}{2} + \frac{1}{2}y. \tag{3}$$

If we solve the equations (2) and (3), we find $x = \frac{2}{3}$ and $y = \frac{1}{3}$. We obtain $z = 0$ from equation (1).

The fact that the probability of the game continuing indefinitely without an outcome is equal to zero can be obtained by direct calculation (exactly as can be done for the probability of the event that no six comes up in rolling a die).

PROBLEM 5. (*a*) We denote by $p_{11}^{(n)}$ the probability that the particle, on starting out from A, is again at A after n moves, and by $p_{12}^{(n)}$ the probability that the particle, on starting out from A, is at B after n moves. If the particle starts out from point A, the event that the particle is at A after $n - 1$ moves and the event that this particle is at B after $n - 1$ moves form a complete system of mutually exclusive events; hence, on the basis of Property 7,

$$p_{11}^{(n)} = p_{11}^{(n-1)}p_{11} + p_{12}^{(n-1)}p_{21}.$$

On the other hand, $p_{11}^{(n-1)} + p_{12}^{(n-1)} = 1$ (Property 3); from this it follows that

$$\begin{aligned}
p_{11}^{(n)} &= p_{11}^{(n-1)}p_{11} + (1 - p_{11}^{(n-1)})p_{21} \\
&= p_{11}^{(n-1)}p_{11} + p_{21} - p_{11}^{(n-1)}p_{21} \\
&= p_{21} + (p_{11} - p_{21})p_{11}^{(n-1)}.
\end{aligned}$$

If we set

$$p_{11} - p_{21} = q,$$

we obtain

$$\begin{aligned}
p_{11}^{(n)} &= p_{21} + qp_{11}^{(n-1)} \\
&= p_{21} + q(p_{21} + qp_{11}^{(n-2)}) \\
&= p_{21} + qp_{21} + q^2 p_{11}^{(n-2)} \\
&= p_{21} + qp_{21} + q^2(p_{21} + qp_{11}^{(n-3)}) \\
&= p_{21} + qp_{21} + q^2 p_{21} + q^3 p_{11}^{(n-3)} \\
&= p_{21} + qp_{21} + q^2 p_{21} + \cdots + q^{n-2}p_{21} + q^{n-1}p_{11}^{[n-(n-1)]}.
\end{aligned}$$

But,

$$p_{11}^{[n-(n-1)]} = p_{11}^{1} = p_{11},$$

from which we obtain

$$p_{11}^{(n)} = p_{21}(1 + q + q^2 + \cdots + q^{n-2}) + q^{n-1}p_{11},$$

or, if we sum by the formula for geometric series,[1]

$$p_{11}^{(n)} = p_{21}\frac{1 - q^{n-1}}{1 - q} + q^{n-1}p_{11} = \frac{p_{21}}{1 - q} + q^{n-1}\left(p_{11} - \frac{p_{21}}{1 - q}\right).$$

By taking into consideration that $p_{11} + p_{12} = 1$ (Property 3), we obtain:

$$1 - q = 1 - p_{11} + p_{21} = p_{12} + p_{21},$$

$$p_{11}^{(n)} = \frac{p_{21}}{p_{12} + p_{21}} + q^{n-1}\left(p_{11} - \frac{p_{21}}{p_{12} + p_{21}}\right)$$

$$= \frac{p_{21}}{p_{12} + p_{21}} + q^{n-1}\frac{p_{12}p_{11} + p_{21}p_{11} - p_{21}}{p_{12} + p_{21}}$$

$$= \frac{p_{21}}{p_{12} + p_{21}} + q^{n-1}\frac{p_{12}p_{11} - p_{21}(1 - p_{11})}{p_{12} + p_{21}}$$

$$= \frac{p_{21}}{p_{12} + p_{21}} + (p_{11} - p_{21})^{n-1}\frac{p_{12}p_{11} - p_{21}p_{12}}{p_{12} + p_{21}}$$

$$= \frac{p_{21}}{p_{12} + p_{21}} + \frac{p_{12}}{p_{12} + p_{21}}(p_{11} - p_{21})^{n}.$$

(b) To find the probability $p_{21}^{(n)}$ that the point A is reached after n moves on starting out from B, we first calculate $p_{12}^{(n)}$. We find this probability by the formula

$$p_{12}^{(n)} = 1 - p_{11}^{(n)}$$

$$= 1 - \frac{p_{21}}{p_{12} + p_{21}} - \frac{p_{12}}{p_{12} + p_{21}}(p_{11} - p_{21})^{n}$$

$$= \frac{p_{12}}{p_{12} + p_{21}} - \frac{p_{12}}{p_{12} + p_{21}}(p_{11} - p_{21})^{n}.$$

The probability $p_{21}^{(n)}$ arises from the probability $p_{12}^{(n)}$ when we exchange the points A and B and, correspondingly, replace the index 1 by 2 and 2 by 1. Hence,

$$p_{21}^{(n)} = \frac{p_{21}}{p_{21} + p_{12}} - \frac{p_{21}}{p_{21} + p_{12}}(p_{22} - p_{12})^{n}.$$

[1] See, for example, I. S. Sominskii, *The Method of Mathematical Induction* (Boston: D. C. Heath and Company, 1963), pages 14 and 40.

Remark. The probabilities $p_{11}^{(n)}$ and $p_{21}^{(n)}$ differ by the quantity

$$p_{11}^{(n)} - p_{21}^{(n)} = \frac{p_{12}}{p_{12} + p_{21}}(p_{11} - p_{21})^n + \frac{p_{21}}{p_{12} + p_{21}}(p_{22} - p_{12})^n. \quad (1)$$

With the exception of the degenerate cases

$$p_{11} = 1, \; p_{12} = 0, \; p_{21} = 0, \; p_{22} = 1,$$
$$p_{11} = 0, \; p_{12} = 1, \; p_{21} = 1, \; p_{22} = 0,$$

we always have that

$$-1 < p_{11} - p_{21} < 1,$$
$$-1 < p_{22} - p_{12} < 1.$$

Hence, as n increases, the summands of the right-hand side of equality (1) approach zero, since they are terms of a decreasing geometric progression. Hence, the difference $p_{11}^{(n)} - p_{21}^{(n)}$ also becomes ever closer to zero. In other words, with increasing n, the probability that the particle is at the point A after n moves becomes increasingly independent of the point from which it starts out.

PROBLEM 6. We shall represent the number of balls in the left urn by means of a marker that moves on the line of numbers. At the beginning, the marker is at point a (Fig. 27). In each unit of time it moves with probability $\frac{1}{2}$ to the

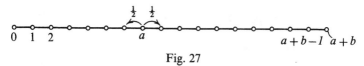

Fig. 27

right (if a ball is moved from the right urn to the left, so that the number of balls in the left urn increases by one), and with the same probability, to the left (if a ball is moved from the left urn to the right, so that the number of balls in the left urn decreases by one). This goes on until the marker reaches the point 0 or the point $(a + b)$ for the first time. If the marker reaches the point 0, the left urn has become empty, and if it reaches the point $a + b$, the right urn has become empty. We denote the probability that the marker reaches the point 0, under the condition that it starts out from the point k, by p_k (p_a is sought). Obviously, $p_0 = 1$ and $p_{a+b} = 0$.

Suppose that the marker is at point k.

We consider the two events:

A: The marker is at $k + 1$ after one move.

B: The marker is at $k - 1$ after one move.

By assumption, $\mathbf{P}(A) = \mathbf{P}(B) = \frac{1}{2}$. If the event A occurs, the probability that the marker reaches the point 0 is equal to p_{k+1}, and if the event B occurs, this probability is equal to p_{k-1}.

By the formula for complete probability, we have

$$p_k = \frac{1}{2}p_{k+1} + \frac{1}{2}p_{k-1},$$

from which we obtain

$$2p_k = p_{k+1} + p_{k-1}, \qquad p_{k+1} - p_k = p_k - p_{k-1}.$$

We denote the constant difference $p_{k+1} - p_k$ by d and write

$$p_k - p_{k-1} = d,$$
$$p_{k-1} - p_{k-2} = d,$$
$$\cdots\cdots\cdots\cdots$$
$$p_2 - p_1 = d,$$
$$p_1 - p_0 = d.$$

By addition of these equalities we obtain

$$p_k - p_0 = kd,$$
$$p_k - 1 = kd,$$
$$p_k = 1 + kd,$$

or, setting $k = a + b$,

$$0 = p_{a+b} = 1 + (a + b)d, \quad d = -\frac{1}{a+b},$$
$$p_k = 1 - \frac{k}{a+b} = \frac{a+b-k}{a+b}.$$

The probability in which we are interested equals

$$p_a = \frac{b}{a+b}.$$

The probability that the right urn becomes empty is then, obviously,

$$p_b = \frac{a}{a+b}.$$

The probability that the experiment ends, that is, that one of the urns becomes empty, is, by Property 2,

$$\frac{b}{a+b} + \frac{a}{a+b} = 1.$$

The probability that the experiment never ends is, by Property 3,

$$1 - 1 = 0$$

Remark. This problem is known in the history of mathematics as the "gambler's ruin problem." Its classical formulation is as follows:

Two people gamble. The probability of a victory for each of them during each game is equal to $\frac{1}{2}$. The first gambler has a rubles, the second b rubles. Play continues until one of the players loses his last ruble. What is the probability of ruin for each of the gamblers?

In Problem 6, the players are replaced by urns and the money by balls.

PROBLEM 7. At each vertex of the cube we write the probability (Fig. 28) that the caterpillar after leaving this vertex gets stuck at the point A. The same probability x is written at the vertices 1 and 5 of Figure 5, since these vertices have the same position with respect to the points A and B. Likewise, the same probability y is written at the vertices 2 and 4. We denote the probability that the caterpillar crawls to A from the points 0 and 3 by z and u, respectively.

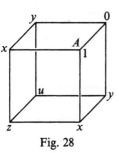

Fig. 28

We now assume that the caterpillar is at the vertex 1 (Fig. 5), and consider the complete system of pairwise mutually exclusive events:

C_1: The caterpillar crawls along the path $1A$,

$$\mathbf{P}(C_1) = \frac{1}{3}.$$

C_2: The caterpillar crawls along the path 12,

$$\mathbf{P}(C_2) = \frac{1}{3}.$$

C_3: The caterpillar crawls along the path 10,

$$\mathbf{P}(C_3) = \frac{1}{3}.$$

The probability that the caterpillar reaches the point A is equal to 1 under the condition C_1, equal to y under the condition C_2, and equal to z under the condition C_3. Hence, (by Property 7),

$$x = \frac{1}{3} \cdot 1 + \frac{1}{3}y + \frac{1}{3}z. \qquad (1)$$

Similarly, one finds the three relations

$$y = \frac{1}{3} \cdot 0 + \frac{1}{3}x + \frac{1}{3}u, \qquad (2)$$

$$z = \frac{1}{3}x + \frac{1}{3}x + \frac{1}{3}u, \qquad (3)$$

$$u = \frac{1}{3}y + \frac{1}{3}y + \frac{1}{3}z. \qquad (4)$$

We find

$$x = \frac{9}{14}, \ y = \frac{5}{14}, \ z = \frac{4}{7}, \ u = \frac{3}{7},$$

as solutions of the equations (1), (2), (3), and (4).

The probability z' that the point B is reached if the caterpillar leaves from the point 0 is found by the following considerations. Obviously, the probability that B is reached from the vertex 0 is equal to the probability that the point A is reached from the vertex 3. Hence, $z' = u = \frac{3}{7}$. By the same considerations, we find

$$x' = y = \frac{5}{14}, \; y' = x = \frac{9}{14}, \; u' = z = \frac{4}{7}.$$

Remark. We remark that

$$x + x' = y + y' = z + z' = u + u' = 1.$$

But, by Property 2, $z + z'$ is the probability that the caterpillar reaches either point starting from the vertex 0, and $1 - (z + z')$ is the probability that the caterpillar reaches neither point starting from the same vertex (Property 3).

Thus, the probability that the caterpillar reaches neither point A nor point B on starting out from the point 0 is equal to zero. One easily sees that it is equally true when any other vertex is chosen as the starting point.

We have already met a similar result in the solution of Problem 6, where it was shown that the probability that the marker reaches neither 0 nor $a + b$ is equal to zero. The probability that the game of Problem 4 never ends is also equal to zero. Finally one can easily ascertain that in the diagram of Problem 5, the probability that the particle never reaches, say, point B is likewise equal to zero (if $p_{12} > 0$).

All these facts are special cases of the general theorem about random walks, which we prove in Chapter 2.

PROBLEM 8. We consider the events:

A: After n moves, the marker is at point k.
B_1: After $n - 1$ moves, the marker is at point $k - 1$.
B_2: After $n - 1$ moves, the marker is at the point $k + 1$.
B_3: After $n - 1$ moves, the marker is neither at $k - 1$ nor at $k + 1$.

By Property 7,

$$\mathbf{P}(A) = \mathbf{P}(B_1)\,\mathbf{P}(A|B_1) + \mathbf{P}(B_2)\,\mathbf{P}(A|B_2) + \mathbf{P}(B_3)\,\mathbf{P}(A|B_3),$$

and we have,

$$\mathbf{P}(A) = Z_n{}^k, \; \mathbf{P}(B_1) = Z_{n-1}{}^{k-1}, \; \mathbf{P}(B_2) = Z_{n-1}{}^{k+1},$$

$$\mathbf{P}(A|B_1) = \frac{1}{2}, \; \mathbf{P}(A|B_2) = \frac{1}{2}, \; \mathbf{P}(A|B_3) = 0.$$

From this, it follows that

$$Z_n{}^k = \frac{Z_{n-1}{}^{k-1} + Z_{n-1}{}^{k+1}}{2}.$$

PROBLEM 9. The elements that stand in the rows of the triangle of probabilities are the probabilities of events that are pairwise mutually exclusive and form a complete system.

PROBLEM 10. For this proof, we write three products, one under the other (compare with (5), (6), and (7) on page 22):

$$\frac{1}{2}\cdot\frac{1}{2}\cdot\frac{3}{4}\cdot\frac{3}{4}\cdot\ldots\cdot\frac{2a-3}{2a-2}\cdot\frac{2a-3}{2a-2}\cdot\frac{2a-1}{2a}\cdot\frac{2a-1}{2a}\cdot\frac{2a}{2a+1}$$

$$\times\frac{2a+1}{2a+2}\cdot\ldots\cdot\frac{2k-2}{2k-1}\cdot\frac{2k-1}{2k},$$

$$\frac{1}{2}\cdot\frac{1}{2}\cdot\frac{3}{4}\cdot\frac{3}{4}\cdot\ldots\cdot\frac{2a-3}{2a-2}\cdot\frac{2a-3}{2a-2}\cdot\frac{2a-1}{2a}\cdot\frac{2a-1}{2a}\cdot\frac{2a+1}{2a+2}$$

$$\times\frac{2a+1}{2a+2}\cdot\ldots\cdot\frac{2k-1}{2k}\cdot\frac{2k-1}{2k},$$

$$\frac{1}{2}\cdot\frac{1}{2}\cdot\frac{3}{4}\cdot\frac{3}{4}\cdot\ldots\cdot\frac{2a-3}{2a-2}\cdot\frac{2a-3}{2a-2}\cdot\frac{2a-1}{2a}\cdot\frac{2a}{2a+1}\cdot\frac{2a+1}{2a+2}$$

$$\times\frac{2a+2}{2a+3}\cdot\ldots\cdot\frac{2k-1}{2k}\cdot\frac{2k}{2k+1}.$$

We see, immediately, that the second product is not smaller than the first, and the third is greater than the second. However, the second product is equal to w_{2k}^2, and, after a few simplifications, the first and third take the form

$$\left(\frac{1}{2}\cdot\frac{3}{4}\cdot\ldots\cdot\frac{2a-3}{2a-2}\right)^2\cdot\frac{2a-1}{2a}\cdot\frac{2a-1}{2a}\cdot\frac{2a}{2a+1}\cdot\frac{2a+1}{2a+2}\cdot\ldots\cdot\frac{2k-1}{2k}$$

$$=\left(\frac{1}{2}\cdot\frac{3}{4}\cdot\ldots\cdot\frac{2a-3}{2a-2}\right)^2\cdot\frac{2a-1}{2a}\cdot\frac{2a-1}{2k},$$

$$\left(\frac{1}{2}\cdot\frac{3}{4}\cdot\ldots\cdot\frac{2a-3}{2a-2}\right)^2\cdot\frac{2a-1}{2a}\cdot\frac{2a}{2a+1}\cdot\frac{2a+1}{2a+2}\cdot\frac{2a+2}{2a+3}\cdot\ldots\cdot\frac{2k}{2k+1}$$

$$=\left(\frac{1}{2}\cdot\frac{3}{4}\cdot\ldots\cdot\frac{2a-3}{2a-2}\right)^2\cdot\frac{2a-1}{2k+1}.$$

Hence,

$$(2a-1)\left(\frac{1}{2}\cdot\frac{3}{4}\cdot\ldots\cdot\frac{2a-3}{2a-2}\right)^2\frac{2a-1}{2a}\frac{1}{2k}$$

$$\leq w_{2k}^2 < (2a-1)\cdot\left(\frac{1}{2}\cdot\frac{3}{4}\cdot\ldots\cdot\frac{2a-3}{2a-2}\right)^2\cdot\frac{1}{2k},$$

and the desired inequality follows, on taking the square root.

PROBLEM 11. For $k \geq 150$ (inequality (16) on page 24),

$$\frac{1}{\sqrt{3.15k}} \leq w_{2k} < \frac{1}{\sqrt{3.14k}},$$

from which it follows that

$$\frac{1}{1.7749\sqrt{k}} < w_{2k} < \frac{1}{1.7719\sqrt{k}},$$

$$\frac{0.5633}{\sqrt{k}} < w_{2k} < \frac{0.5644}{\sqrt{k}}.$$

In our case, $k = \dfrac{10,000}{2} = 5,000$, and, hence,

$$70.710 < \sqrt{k} < 70.711; \quad 0.01414 < \frac{1}{\sqrt{k}} < 0.01415.$$

Finally, we obtain

$$0.5633 \times 0.01414 < w_{2k} < 0.5644 \times 0.01415,$$
$$0.007965 < w_{2k} < 0.007987.$$

On rounding off this value to two significant digits, we have

$$w_{2k} = 0.0080.$$

PROBLEM 12. Let $n = 2k$. All terms of the nth row are smaller than the middle term, which satisfies the inequality

$$w_n = w_{2k} < \frac{1}{\sqrt{2k}} = \frac{1}{\sqrt{n}}.$$

Now, let $n = 2k - 1$. There is no middle term in odd-numbered rows, and the largest terms in these rows are the equal terms Z_n^{-1} and Z_n^{1}. But,

$$Z_n^{1} = \frac{Z_n^{-1} + Z_n^{1}}{2} = Z_{n+1}^{0}.$$

As just shown,

$$Z_{n+1}^{0} < \frac{1}{\sqrt{n+1}}.$$

Hence,

$$Z_n^{1} < \frac{1}{\sqrt{n+1}} < \frac{1}{\sqrt{n}};$$

the other terms of the nth row are thus certainly smaller than $\dfrac{1}{\sqrt{n}}$.

PROBLEM 13. Since the sth term from the middle term of the $2k$th row is $Z_{2k}{}^{2s}$, one can rewrite the inequality (19) on page 26 in the following way:

$$\left(\frac{k+1-s}{k+1}\right)^s \le \frac{Z_{2k}{}^{2s}}{w_{2k}} \le \left(\frac{k}{k+s}\right)^s,$$

or,

$$w_{2k}\left(\frac{k+1-s}{k+1}\right)^s \le Z_{2k}{}^{2s} \le w_{2k}\left(\frac{k}{k+s}\right)^s.$$

For $k \ge 60$ (see inequality (15) on page 24),

$$\frac{1}{\sqrt{3.18k}} \le w_{2k} < \frac{1}{\sqrt{3.10k}}.$$

Hence, for $k \ge 60$,

$$\frac{1}{\sqrt{3.18k}}\left(\frac{k+1-s}{k+1}\right)^s \le Z_{2k}{}^{2s} \le \frac{1}{\sqrt{3.10k}}\left(\frac{k}{k+s}\right)^s.$$

For an approximation to $Z_{120}{}^{20}$, one must set $s = 10$ and $k = 60$. Then,

$$\frac{1}{\sqrt{3.18 \cdot 60}}\left(\frac{51}{61}\right)^{10} \le Z_{120}{}^{20} \le \frac{1}{\sqrt{3.1 \cdot 60}}\left(\frac{60}{70}\right)^{10},$$

and, hence,

$$0.012 < Z_{120}{}^{20} < 0.016.$$

PROBLEM 14. We have

$$(1 + p)^r = \underbrace{(1 + p)(1 + p) \cdots (1 + p)}_{r \text{ times}}.$$

Multiplying out these factors, we obtain a number of terms. One of the terms is 1, obtained from multiplying together the 1 from each factor. After that, we obtain p by multiplying p from the first factor by 1 from all the other factors; we likewise obtain p on taking p from the second factor and multiplying by 1 from all the other parentheses, etc. Hence, p occurs r times in the expansion and

$$(1 + p)^r = 1 + \underbrace{(p + p + \cdots + p)}_{r \text{ terms}} + \cdots$$

$$= 1 + rp + \cdots.$$

The terms we have not written out are all positive; hence,

$$(1 + p)^r \ge 1 + rp.$$

PROBLEM 15. One can take $\dfrac{2}{\sqrt[3]{0.05}} \cdot \sqrt{n} = 5.429 \sqrt{n}$ as the practically certain bound for the displacement after n moves. In particular, for $n = 1{,}000$, 171.7 is the practically certain bound of the displacement. The largest even number that does not exceed 171.7 is 170; hence, one can state with practical certainty that the marker is not more than 170 steps from the starting position after 1,000 moves.

To obtain the practically certain bound for the reduced velocity, one must divide the bound for the deviation by the number of moves. We find that, with practical certainty, the marker has a reduced velocity of at most 0.17 after 1,000 moves.

PROBLEM 16. We set

$$\varepsilon = \left(\frac{2}{\alpha\sqrt{n}}\right)^3.$$

One can say with the probability of error less than ε that after n moves the reduced velocity of the marker is less than $\dfrac{2}{\sqrt[3]{\varepsilon}\sqrt{n}}$. But,

$$\frac{2}{\sqrt[3]{\varepsilon}\sqrt{n}} = \frac{2}{\dfrac{2}{\alpha\sqrt{n}}\sqrt{n}} = \alpha.$$

PROBLEM 17. One can assert with a probability of error less than 0.001 that after n moves the reduced velocity of the marker is less than $\dfrac{2}{\sqrt[3]{0.001}\sqrt{n}}$.

We now choose n in such a way that

$$\frac{2}{\sqrt[3]{0.001}\sqrt{n}} \le 0.01;$$

from this, it follows that $n \ge 4 \cdot 10^6$.

Hence, one can assert with the probability of error less than 0.001 that the reduced velocity after $4 \cdot 10^6$ moves is smaller than

$$\frac{2}{\sqrt[3]{0.001}\sqrt{4 \cdot 10^6}} = 0.01.$$

PROBLEM 18. In order to assert with probability of error less than 0.01 that the frequency with which heads comes up differs from 0.5 by less than 0.1, we can take any number of tosses that is greater than

$$N = \frac{1}{(0.1)^2 \sqrt[3]{0.01}} = 464.17.$$

It is thus sufficient to make 465 tosses.

PROBLEM 19. If the particle is at m after n moves, it could have arrived there first after one move, after two moves, . . . , or after n moves. Of interest to us is the probability that is expressed, with the help of Formula (5) of section 2, as the sum

$$b_1 + b_2 + \cdots + b_n,$$

where b_k is the probability that m is reached for the first time after k moves. Hence, it suffices to show that the probability b_k does not depend on a barrier.

We now prove that b_k does not depend on what kind of barrier is at the point m. We consider all paths of k moves that begin at the starting point, end at the point m, but otherwise do not touch the point m. These paths are denoted by $A_1, A_2, \ldots, A_s$. We denote by a_i the probability that the particle uses the path A_i for its first k moves. The event that the particle reaches the point m for the first time after k moves is the same as the event that it used either the path A_1, the path $A_2, \ldots,$ or, finally, the path A_s. Thus,

$$b_k = a_1 + a_2 + \cdots + a_s.$$

However, none of the paths $A_1, A_2, \ldots, A_s$ goes through m; hence, the probabilities $a_1, a_2, \ldots, a_s$ (and with them also b_k) do not depend on whether or not there is a barrier at the point m or what kind of barrier it may be.

Remark. These considerations have a general character and can be applied to every scheme of a random walk. Hence, the statement formulated in the exercise is valid for arbitrary schemes.

PROBLEM 20. From the relations (1), (2), and (3) on page 38, we find that

$$b_n = \frac{1}{2} b_{n-2} + \frac{1}{4} b_{n-2} = \frac{3}{4} b_{n-2},$$

$$b_{2k} = \left(\frac{3}{4}\right)^k b_0,$$

$$b_{2k+1} = \left(\frac{3}{4}\right)^k b_1.$$

But, $b_0 = 1$ (at the beginning the particle is at 1) and $b_1 = 0$ (after one move, the particle is no longer at 1). Hence,

$$\left. \begin{aligned} b_{2k} &= \left(\frac{3}{4}\right)^k, \\[2mm] b_{2k+1} &= 0. \end{aligned} \right\} \tag{1}$$

It follows from the equalities (3) and (4) on page 38 that

$$d_n = d_{n-1} + \left(\frac{1}{4}\right) b_{n-2}.$$

In addition, by taking into consideration equation (1), we obtain

$$d_{2k+1} = d_{2k} + \frac{1}{4}b_{2k-1} = d_{2k},$$

$$d_{2k} = d_{2k-1} + \frac{1}{4}b_{2k-2} = d_{2k-2} + \frac{1}{4}\left(\frac{3}{4}\right)^{k-1}$$

$$= d_{2k-4} + \frac{1}{4}\left(\frac{3}{4}\right)^{k-2} + \frac{1}{4}\left(\frac{3}{4}\right)^{k-1}$$

$$= d_0 + \frac{1}{4}\left(\frac{3}{4}\right)^{0} + \frac{1}{4}\left(\frac{3}{4}\right)^{1} + \cdots + \frac{1}{4}\left(\frac{3}{4}\right)^{k-1}.$$

Since $d_0 = 0$, we finally obtain (by using the formula for the sum of a geometric series)

$$d_{2k} = \frac{1}{4}\left[1 + \frac{3}{4} + \cdots + \left(\frac{3}{4}\right)^{k-1}\right] = \frac{1}{4} \cdot \frac{1 - (\frac{3}{4})^k}{1 - \frac{3}{4}} = 1 - \left(\frac{3}{4}\right)^{k}.$$

PROBLEM 21.

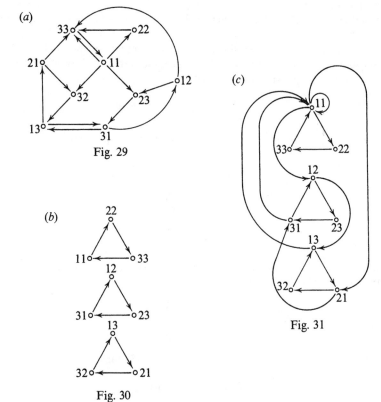

(a)

Fig. 29

(b)

(c)

Fig. 30

Fig. 31

PROBLEM 22. (*a*) The second particle, just as the first, can also reach E_k from E_1 in z moves.[1] If we relate this path of the two particles to the motion of the marker on K^2, we find that the marker reaches E_{kk} from E_{11} in z moves.

(*b*) It suffices to prove that one can reach E_{11} from an arbitrary position E_{ik} of the chain L. (One can, then, also reach any arbitrary position E_{jl} from it along the path $E_{ik} - E_{11} - E_{jl}$.) Thus, we have to show that one can reach E_{11} from the position E_{ik} belonging to L. Suppose that one reaches the point E_{ik} from E_{11} after s moves. In this case, the first particle goes from E_1 to E_i in s moves, and the second goes from E_1 to E_k (Fig. 32). Now, let E_1

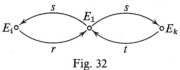

Fig. 32

be reached from E_i in r moves and E_1 from E_k in t moves (the chain K is irreducible!). Then (Fig. 32), the first particle can go from E_i to E_1 in $r + s + t$ moves (on the path $E_i - E_1 - E_k - E_1$) and the second particle from E_k to E_1 in $t + s + r$ moves (on the path $E_k - E_1 - E_i - E_1$); that is, the first particle requires the same number of moves to go from E_i to E_1 as the second does to go from E_k to E_1, namely $z = s + r + t$ moves. If this path of the two particles is carried over to the motion of the marker on K^2, the marker can reach E_{11} from E_{ik} in z moves.

PROBLEM 23. Since all points are fully equivalent, x is really the probability that a particle leaving $n - 1$ reaches n at some time. Likewise, y is the probability that a particle leaving $n + 1$ reaches n.

We assume that the particle starts out from the point 0 and consider the events:

A: The particle returns to the point 0 at some time.
B_1: The particle reaches the point 1 on the first move.
B_2: The particle reaches the point -1 on the first move.

By the formula for complete probability (Property 7, section 3), we have

$$\mathbf{P}(A) = \mathbf{P}(B_1)\,\mathbf{P}(A|B_1) + \mathbf{P}(B_2)\,\mathbf{P}(A|B_2). \qquad (1)$$

If the condition B_1 is fulfilled, the particle is at the point 1 after one move. For the event A to occur under this condition it is necessary that the particle reach 0 at some time on starting out from 1. The probability for this is equal to y. Hence, $\mathbf{P}(A|B_1) = y$. Likewise, $\mathbf{P}(A|B_2) = x$. Furthermore, $\mathbf{P}(A) = z$ and $\mathbf{P}(B_1) = \mathbf{P}(B_2) = \frac{1}{2}$. If these values are substituted in formula (1), we obtain

$$z = \frac{1}{2}y + \frac{1}{2}x. \qquad (2)$$

Furthermore, it is obvious that $x = y$, and, thus, $z = x$.

[1] "Can reach" means "can reach by moving in the direction of the arrows."

To calculate x, we consider the event C (that the particle reaches 1 from 0 at some time). By Property 7, we have

$$x = \mathbf{P}(C) = \mathbf{P}(B_1)\mathbf{P}(C|B_1) + \mathbf{P}(B_2)\mathbf{P}(C|B_2). \tag{3}$$

We have $\mathbf{P}(B_1) = \mathbf{P}(B_2) = \frac{1}{2}$ and $\mathbf{P}(C|B_1) = 1$. Furthermore, $\mathbf{P}(C|B_2)$ is the probability that the particle reaches 1 at some time on starting from -1. We shall calculate this probability. Thus, suppose the particle is located at -1. The probability of the event D (the particle reaches the point 1 at some time) must now be found. For this, we introduce the event F (the particle reaches 0 at some time). Obviously, $FD = D$ (for, the occurrence of D is equivalent to the joint occurrence of the events F and D). Hence, by Property 6,

$$\mathbf{P}(D) = \mathbf{P}(FD) = \mathbf{P}(F)\mathbf{P}(D|F).$$

But, $\mathbf{P}(F) = x$, $\mathbf{P}(D|F) = x$. Hence,

$$\mathbf{P}(D) = x^2. \tag{4}$$

Finally, we find from the formulas (3) and (4) that:

$$x = \frac{1}{2} + \frac{1}{2}x^2, \tag{5}$$

$$x^2 - 2x + 1 = 0,$$

$$(x - 1)^2 = 0,$$

from which it follows that $x = 1$, and, thus, $y = z = 1$.

PROBLEM 24. It must be shown that the particle reaches the point n with probability 1 (for the sake of definiteness, let $n > 0$). For $n = 1$, this was proved in the previous exercise. Let the statement be proved for the point n; we will prove that it then holds for $n + 1$ as well.

Let us consider the event A_{n+1} (the particle reaches $n + 1$ at some time). Obviously, $A_{n+1} = A_n A_{n+1}$; by Property 6, we then have

$$\mathbf{P}(A_{n+1}) = \mathbf{P}(A_n)\mathbf{P}(A_{n+1}|A_n).$$

But, $\mathbf{P}(A_{n+1}|A_n) = x = 1$. Furthermore, $\mathbf{P}(A_n) = 1$, by assumption. Hence, $\mathbf{P}(A_{n+1}) = 1$.

The probability that a preassigned point is reached twice, three times, or, in general, n times, is found exactly as in Remark 1 on page 43; it is equal to 1.

PROBLEM 25. We remark first that $p + p + r = 1$. (See footnote 1 on page 46.) We may denote by x, y, z the probabilities of the same events as in Problem 23. We leave it to the reader to find for himself the equations,

$$z = r + px + py, \tag{1}$$

$$x = p + rx + px^2. \tag{2}$$

(One derives them exactly as (2) and (5) in the solution of Problem 23. One has merely to consider, in addition to B_1 and B_2, the event B_3: the particle remains at the point 0 after the first move). Obviously, $x = y$. Hence,

$$z = r + 2px.$$

We transform (2):

$$x = p + (1 - 2p)x + px^2,$$
$$0 = p - 2px + px^2.$$

Since $p \neq 0$,

$$1 - 2x + x^2 = 0,$$

from which one obtains $x = 1$ and $z = r + 2p = 1$.

Exactly as in Problem 24, we ascertain that the particle reaches every point with the probability 1.

PROBLEM 26. With the same considerations as were applied in the solution of Problems 23 and 25, we obtain

$$z = py + qx, \tag{1}$$

$$x = p + qx^2, \tag{2}$$

$$y = q + py^2 \tag{3}$$

where, in general, $x \neq y$. The equation (2) yields

$$x = \frac{1 \pm \sqrt{1 - 4pq}}{2q}$$

$$= \frac{1 \pm \sqrt{1 - 4p(1 - p)}}{2q}$$

$$= \frac{1 \pm \sqrt{(1 - 2p)^2}}{2q}$$

$$= \frac{1 \pm (1 - 2p)}{2q}$$

as solutions; the equation (2) thus has the roots

$$x_1 = \frac{1 + 1 - 2p}{2q} = \frac{2 - 2p}{2q} = \frac{1 - p}{q} = \frac{q}{q} = 1,$$

$$x_2 = \frac{1 - 1 + 2p}{2q} = \frac{2p}{2q} = \frac{p}{q}.$$

Since $\frac{p}{q} > 1$ and x (like every probability) is less than or equal to 1, $x \neq x_2$ and, hence, $x = x_1 = 1$.

We obtain

$$y_1 = 1, \quad y_2 = \frac{q}{p}$$

as a solution of equation (3). By assumption, $y \neq 1$, and, therefore, $y = \frac{q}{p}$. Hence,

$$z = py + qx = 2q.$$

PROBLEM 27. We prove that for $k \geq 2^m$ the inequality $S_k > \frac{m}{2}$ is satisfied. We have

$$S_{2^m} = 1 + \frac{1}{2} + \frac{1}{3} + \cdots + \frac{1}{2^m} = 1 + T_1 + T_2 + \cdots + T_m,$$

where

$$T_p = \underbrace{\frac{1}{2^{p-1} + 1} + \frac{1}{2^{p-1} + 2} + \cdots + \frac{1}{2^p}}_{2^{p-1} \text{ summands}}.$$

Clearly,

$$T_p \geq \underbrace{\frac{1}{2^p} + \frac{1}{2^p} + \cdots + \frac{1}{2^p}}_{2^{p-1} \text{ summands}} = \frac{2^{p-1}}{2^p} = \frac{1}{2}.$$

Hence,

$$S_k \geq S_{2^m} \geq 1 + \underbrace{\frac{1}{2} + \frac{1}{2} + \cdots + \frac{1}{2}}_{m \text{ summands}} = 1 + \frac{m}{2}.$$

PROBLEM 28. We consider the straight lines N_1N_1, $N_{-1}N_{-1}$, N_2N_2, $N_{-2}N_{-2}$, that run parallel to the line NN (Fig. 26). Let a traveler be at a point of the line N_kN_k. The traveler goes upward on the next move with a probability $\frac{1}{4}$ and downward on the next move with a probability $\frac{1}{4}$; that is, he remains on the line N_kN_k with the probability $\frac{1}{4} + \frac{1}{4} = \frac{1}{2}$. The traveler goes to the right and reaches the line $N_{k+1}N_{k+1}$ with a probability $\frac{1}{4}$ or goes to the left and reaches the line $N_{k-1}N_{k-1}$ with the probability $\frac{1}{4}$. If we regard the line N_kN_k as one position, we obtain the situation that we have already dealt with in Problem 25. The lines N_kN_k here correspond to the points, with $p = \frac{1}{4}$ and $r = \frac{1}{2}$. By what was proved there, the traveler reaches each line with probability 1; thus, also the line NN.

A CATALOG OF SELECTED
DOVER BOOKS
IN SCIENCE AND MATHEMATICS

Mathematics

FUNCTIONAL ANALYSIS (Second Corrected Edition), George Bachman and Lawrence Narici. Excellent treatment of subject geared toward students with background in linear algebra, advanced calculus, physics and engineering. Text covers introduction to inner-product spaces, normed, metric spaces, and topological spaces; complete orthonormal sets, the Hahn-Banach Theorem and its consequences, and many other related subjects. 1966 ed. 544pp. 6⅛ x 9¼. 0-486-40251-7

ASYMPTOTIC EXPANSIONS OF INTEGRALS, Norman Bleistein & Richard A. Handelsman. Best introduction to important field with applications in a variety of scientific disciplines. New preface. Problems. Diagrams. Tables. Bibliography. Index. 448pp. 5⅜ x 8½. 0-486-65082-0

VECTOR AND TENSOR ANALYSIS WITH APPLICATIONS, A. I. Borisenko and I. E. Tarapov. Concise introduction. Worked-out problems, solutions, exercises. 257pp. 5⅝ x 8¼. 0-486-63833-2

AN INTRODUCTION TO ORDINARY DIFFERENTIAL EQUATIONS, Earl A. Coddington. A thorough and systematic first course in elementary differential equations for undergraduates in mathematics and science, with many exercises and problems (with answers). Index. 304pp. 5⅜ x 8½. 0-486-65942-9

FOURIER SERIES AND ORTHOGONAL FUNCTIONS, Harry F. Davis. An incisive text combining theory and practical example to introduce Fourier series, orthogonal functions and applications of the Fourier method to boundary-value problems. 570 exercises. Answers and notes. 416pp. 5⅜ x 8½. 0-486-65973-9

COMPUTABILITY AND UNSOLVABILITY, Martin Davis. Classic graduate-level introduction to theory of computability, usually referred to as theory of recurrent functions. New preface and appendix. 288pp. 5⅜ x 8½. 0-486-61471-9

ASYMPTOTIC METHODS IN ANALYSIS, N. G. de Bruijn. An inexpensive, comprehensive guide to asymptotic methods–the pioneering work that teaches by explaining worked examples in detail. Index. 224pp. 5⅜ x 8½ 0-486-64221-6

APPLIED COMPLEX VARIABLES, John W. Dettman. Step-by-step coverage of fundamentals of analytic function theory–plus lucid exposition of five important applications: Potential Theory; Ordinary Differential Equations; Fourier Transforms; Laplace Transforms; Asymptotic Expansions. 66 figures. Exercises at chapter ends. 512pp. 5⅜ x 8½. 0-486-64670-X

INTRODUCTION TO LINEAR ALGEBRA AND DIFFERENTIAL EQUATIONS, John W. Dettman. Excellent text covers complex numbers, determinants, orthonormal bases, Laplace transforms, much more. Exercises with solutions. Undergraduate level. 416pp. 5⅜ x 8½. 0-486-65191-6

RIEMANN'S ZETA FUNCTION, H. M. Edwards. Superb, high-level study of landmark 1859 publication entitled "On the Number of Primes Less Than a Given Magnitude" traces developments in mathematical theory that it inspired. xiv+315pp. 5⅜ x 8½. 0-486-41740-9

CALCULUS OF VARIATIONS WITH APPLICATIONS, George M. Ewing. Applications-oriented introduction to variational theory develops insight and promotes understanding of specialized books, research papers. Suitable for advanced undergraduate/graduate students as primary, supplementary text. 352pp. 5⅜ x 8½.
0-486-64856-7

COMPLEX VARIABLES, Francis J. Flanigan. Unusual approach, delaying complex algebra till harmonic functions have been analyzed from real variable viewpoint. Includes problems with answers. 364pp. 5⅜ x 8½. 0-486-61388-7

AN INTRODUCTION TO THE CALCULUS OF VARIATIONS, Charles Fox. Graduate-level text covers variations of an integral, isoperimetrical problems, least action, special relativity, approximations, more. References. 279pp. 5⅜ x 8½.
0-486-65499-0

COUNTEREXAMPLES IN ANALYSIS, Bernard R. Gelbaum and John M. H. Olmsted. These counterexamples deal mostly with the part of analysis known as "real variables." The first half covers the real number system, and the second half encompasses higher dimensions. 1962 edition. xxiv+198pp. 5⅜ x 8½. 0-486-42875-3

CATASTROPHE THEORY FOR SCIENTISTS AND ENGINEERS, Robert Gilmore. Advanced-level treatment describes mathematics of theory grounded in the work of Poincaré, R. Thom, other mathematicians. Also important applications to problems in mathematics, physics, chemistry and engineering. 1981 edition. References. 28 tables. 397 black-and-white illustrations. xvii + 666pp. 6⅛ x 9¼.
0-486-67539-4

INTRODUCTION TO DIFFERENCE EQUATIONS, Samuel Goldberg. Exceptionally clear exposition of important discipline with applications to sociology, psychology, economics. Many illustrative examples; over 250 problems. 260pp. 5⅜ x 8½.
0-486-65084-7

NUMERICAL METHODS FOR SCIENTISTS AND ENGINEERS, Richard Hamming. Classic text stresses frequency approach in coverage of algorithms, polynomial approximation, Fourier approximation, exponential approximation, other topics. Revised and enlarged 2nd edition. 721pp. 5⅜ x 8½. 0-486-65241-6

INTRODUCTION TO NUMERICAL ANALYSIS (2nd Edition), F. B. Hildebrand. Classic, fundamental treatment covers computation, approximation, interpolation, numerical differentiation and integration, other topics. 150 new problems. 669pp. 5⅜ x 8½. 0-486-65363-3

THREE PEARLS OF NUMBER THEORY, A. Y. Khinchin. Three compelling puzzles require proof of a basic law governing the world of numbers. Challenges concern van der Waerden's theorem, the Landau-Schnirelmann hypothesis and Mann's theorem, and a solution to Waring's problem. Solutions included. 64pp. 5⅜ x 8½.
0-486-40026-3

THE PHILOSOPHY OF MATHEMATICS: AN INTRODUCTORY ESSAY, Stephan Körner. Surveys the views of Plato, Aristotle, Leibniz & Kant concerning propositions and theories of applied and pure mathematics. Introduction. Two appendices. Index. 198pp. 5⅜ x 8½. 0-486-25048-2

INTRODUCTORY REAL ANALYSIS, A.N. Kolmogorov, S. V. Fomin. Translated by Richard A. Silverman. Self-contained, evenly paced introduction to real and functional analysis. Some 350 problems. 403pp. 5⅜ x 8½. 0-486-61226-0

APPLIED ANALYSIS, Cornelius Lanczos. Classic work on analysis and design of finite processes for approximating solution of analytical problems. Algebraic equations, matrices, harmonic analysis, quadrature methods, much more. 559pp. 5⅜ x 8½. 0-486-65656-X

AN INTRODUCTION TO ALGEBRAIC STRUCTURES, Joseph Landin. Superb self-contained text covers "abstract algebra": sets and numbers, theory of groups, theory of rings, much more. Numerous well-chosen examples, exercises. 247pp. 5⅜ x 8½. 0-486-65940-2

QUALITATIVE THEORY OF DIFFERENTIAL EQUATIONS, V. V. Nemytskii and V.V. Stepanov. Classic graduate-level text by two prominent Soviet mathematicians covers classical differential equations as well as topological dynamics and ergodic theory. Bibliographies. 523pp. 5⅜ x 8½. 0-486-65954-2

THEORY OF MATRICES, Sam Perlis. Outstanding text covering rank, nonsingularity and inverses in connection with the development of canonical matrices under the relation of equivalence, and without the intervention of determinants. Includes exercises. 237pp. 5⅜ x 8½. 0-486-66810-X

INTRODUCTION TO ANALYSIS, Maxwell Rosenlicht. Unusually clear, accessible coverage of set theory, real number system, metric spaces, continuous functions, Riemann integration, multiple integrals, more. Wide range of problems. Undergraduate level. Bibliography. 254pp. 5⅜ x 8½. 0-486-65038-3

MODERN NONLINEAR EQUATIONS, Thomas L. Saaty. Emphasizes practical solution of problems; covers seven types of equations. ". . . a welcome contribution to the existing literature...."–*Math Reviews*. 490pp. 5⅜ x 8½. 0-486-64232-1

MATRICES AND LINEAR ALGEBRA, Hans Schneider and George Phillip Barker. Basic textbook covers theory of matrices and its applications to systems of linear equations and related topics such as determinants, eigenvalues and differential equations. Numerous exercises. 432pp. 5⅜ x 8½. 0-486-66014-1

LINEAR ALGEBRA, Georgi E. Shilov. Determinants, linear spaces, matrix algebras, similar topics. For advanced undergraduates, graduates. Silverman translation. 387pp. 5⅜ x 8½. 0-486-63518-X

ELEMENTS OF REAL ANALYSIS, David A. Sprecher. Classic text covers fundamental concepts, real number system, point sets, functions of a real variable, Fourier series, much more. Over 500 exercises. 352pp. 5⅜ x 8½. 0-486-65385-4

SET THEORY AND LOGIC, Robert R. Stoll. Lucid introduction to unified theory of mathematical concepts. Set theory and logic seen as tools for conceptual understanding of real number system. 496pp. 5⅜ x 8¼. 0-486-63829-4

TENSOR CALCULUS, J.L. Synge and A. Schild. Widely used introductory text covers spaces and tensors, basic operations in Riemannian space, non-Riemannian spaces, etc. 324pp. 5⅜ x 8¼. 0-486-63612-7

ORDINARY DIFFERENTIAL EQUATIONS, Morris Tenenbaum and Harry Pollard. Exhaustive survey of ordinary differential equations for undergraduates in mathematics, engineering, science. Thorough analysis of theorems. Diagrams. Bibliography. Index. 818pp. 5⅜ x 8½. 0-486-64940-7

INTEGRAL EQUATIONS, F. G. Tricomi. Authoritative, well-written treatment of extremely useful mathematical tool with wide applications. Volterra Equations, Fredholm Equations, much more. Advanced undergraduate to graduate level. Exercises. Bibliography. 238pp. 5⅜ x 8½. 0-486-64828-1

FOURIER SERIES, Georgi P. Tolstov. Translated by Richard A. Silverman. A valuable addition to the literature on the subject, moving clearly from subject to subject and theorem to theorem. 107 problems, answers. 336pp. 5⅜ x 8½. 0-486-63317-9

INTRODUCTION TO MATHEMATICAL THINKING, Friedrich Waismann. Examinations of arithmetic, geometry, and theory of integers; rational and natural numbers; complete induction; limit and point of accumulation; remarkable curves; complex and hypercomplex numbers, more. 1959 ed. 27 figures. xii+260pp. 5⅜ x 8½. 0-486-63317-9

POPULAR LECTURES ON MATHEMATICAL LOGIC, Hao Wang. Noted logician's lucid treatment of historical developments, set theory, model theory, recursion theory and constructivism, proof theory, more. 3 appendixes. Bibliography. 1981 edition. ix + 283pp. 5⅜ x 8½. 0-486-67632-3

CALCULUS OF VARIATIONS, Robert Weinstock. Basic introduction covering isoperimetric problems, theory of elasticity, quantum mechanics, electrostatics, etc. Exercises throughout. 326pp. 5⅜ x 8½. 0-486-63069-2

THE CONTINUUM: A CRITICAL EXAMINATION OF THE FOUNDATION OF ANALYSIS, Hermann Weyl. Classic of 20th-century foundational research deals with the conceptual problem posed by the continuum. 156pp. 5⅜ x 8½. 0-486-67982-9

CHALLENGING MATHEMATICAL PROBLEMS WITH ELEMENTARY SOLUTIONS, A. M. Yaglom and I. M. Yaglom. Over 170 challenging problems on probability theory, combinatorial analysis, points and lines, topology, convex polygons, many other topics. Solutions. Total of 445pp. 5⅜ x 8½. Two-vol. set.
Vol. I: 0-486-65536-9 Vol. II: 0-486-65537-7

Paperbound unless otherwise indicated. Available at your book dealer, online at **www.doverpublications.com**, or by writing to Dept. GI, Dover Publications, Inc., 31 East 2nd Street, Mineola, NY 11501. For current price information or for free catalogues (please indicate field of interest), write to Dover Publications or log on to **www.doverpublications.com** and see every Dover book in print. Dover publishes more than 500 books each year on science, elementary and advanced mathematics, biology, music, art, literary history, social sciences, and other areas.